MICROELECTRONICS

Wireless Technology and MEMS in the Developing Countries

MICROELECTRONICS

Wireless Technology and MEMS in the Developing Countries

Editor
Dr. V.K. Jain
Emeritus Scientist
Central Scientific Instrument Organisation
C.S.I.R., New Delhi

CENTRE FOR SCIENCE & TECHNOLOGY OF THE
NON-ALIGNED AND OTHER DEVELOPING COUNTRIES
(NAM S&T CENTRE)

2007
DAYA PUBLISHING HOUSE
Delhi - 110 035

ISBN 81-7035-444-7

Centre for Science and Technology of the Non-Aligned and Other Developing Countries (NAM S&T Centre)
Core-6A, 2nd Floor, India Habitat Centre, Lodhi Road,
New Delhi-110 003 (India)
Phone: +91-11-24644974, 24645134, Fax: +91-11-24644973
E-mail: namstct@vsnl.com
Website: www.namstct.org

Published by : **Daya Publishing House**
1123/74, Deva Ram Park, Tri Nagar, Delhi - 110 035
Phone: 27383999, Fax: (011) 23244987
e-mail: dayabooks@vsnl.com
website: www.dayabooks.com
Showroom : 4760-61/23, Ansari Road, Darya Ganj
New Delhi - 110 002
Phone: 23245578, 23244987

Laser Typesetting : **Classic Computer Services**
Delhi - 110 035

Printed at : **Chawla Offset Printers**
Delhi - 110 052

Printed in India

Foreword

Since the 1960s, technological advances in microelectronics have had far-reaching impacts to a broad range of productions and services sectors, and innovations have transformed many sectors of the global economy. Hot emerging technologies of today - wireless, nano-technology, micro-electro-mechanical systems (MEMS) and robotics, continue to radically improve the quality of life and has become a vital economic force to some developed countries. Recognizing the commercial scientific and the technological capabilities, MEMS and wireless have been identified as enabling technologies that have the potential to significantly enhance Malaysia's technology directions.

MIMOS, as the lead institution in microelectronic research and technology development, with the support from Ministry of Science, Technology and Innovation (MOSTI) contribute significant role in the development of the semiconductor industry in Malaysia. Intensive research activities are being carried out on microelectronic in general, and in MEMS and wireless in particular.

I am very glad that international workshop on microelectronics organized in Kuala Lumpur, Malaysia by Centre for Science and Technology of the Non-Aligned and Other Developing Countries (NAM S&T Centre) jointly with MIMOS between 6-7 September, 2005 has provided excellent opportunity to generate mutually beneficial and effective partnership between the NAM member countries especially on emerging technologies such as of MEMS and wireless. The meeting of minds provides the seedbed for the cross-fertilization of ideas that will be a valuable input in the planning and development of strategies for strengthening the scientific and technological foundation in each participating NAM member countries.

On behalf of the Secretariat on the 5th International Workshop on Microelectronics, I would like to place on record my appreciation and thanks to all the members and respective authors who have contributed to this compendium. I am sure these efforts serve as a significant step towards economic and social well being of the NAM countries. Lastly, I would like to thank government of Malaysia through MOSTI and NAM S&T Centre for providing funding and technical support for making the workshop and this compendium possible.

Wan Mohd Salleh Wan Abu Bakar
Vice President
MIMOS Berhad

From the Editor

Human body is the perfect example of a smart system, having sensors, actuators and brain as intelligent control. If we consider the same system at micro scale, such as a mosquito or an insect possessing similar elements, then a machine developed of the insect size and having similar functions would be called as MEMS. MEMS are intelligent micro-machines having sensors, actuators and signal processing electronics gathered at one place. Known by different names, such as smart systems, micro systems etc., these are non-biological systems but analogous to biological systems. MEMS technology has revolutionized every field by bringing together silicon based microelectronics and micro-machining technologies. It is the general impression that MEMS technology will transform our present day life in a similar way as computers did to us in the last decade. The next revolution in silicon technology would be in the direction of the development of MEMS and this would be different from the previous transformation, which aimed at only the packaging of more transistors onto the silicon chip. The power of the chips will not be restricted to merely storing and processing the information. Currently the incorporation of new type of functionality on the chip is making the structures to sense, act and communicate as well. This is making a chip to have ability to acquire information from the surroundings and communicate what it knows.

Bulk and surface micro machining of silicon has made it possible to decrease the thickness of the structure to a few microns, or even less than a micron, and give the required shape to the device. The devices made of such silicon are very small in size and also of low cost as the devices can be batch fabricated in thousands at a time. These machines have the capability to perform a variety of functions like physical and chemical sensing, analyzing the information and communication. In these devices inertia and gravity are less important rather than the effects of atomic forces surface science. Such devices have applications in every field, but major areas in which one envisages great applications are aerospace, defense, automobiles, biomedical and communications and various other sectors. Bio-MEMS are making revolution in biomedical investigations and clinical use. Presently the investigations are concentrated in the field of catheters, blood pressure measurements and blood analysis. Advanced applications of Bio-MEMS are in Neurology and Cardiology. New investigations are also under development, for example, to make micro-needles to collect the very small quantity of blood from the human body without any pain and feel like a mosquito bite.

Efforts are also on to make a complete lab on a chip. Micro-cantilever is the latest in the bio-detection system. When a molecule gets absorbed at its surface, the micro-cantilever beam bends due to the change in the mass, which can be measured optically or detected through the change in the resonance frequency. These micro-cantilevers can be made using bulk as well as surface micro-machining techniques. In some bio-MEMS systems, it is also important to transfer the data by using wireless system.

These sensors can be divided in to two groups. The first of these has large volume applications linked to automotive and consumer markets. The second group, which is of medium volume, is required for applications such as those in defense, aerospace and medical areas. Presently the production is highly concentrated for the MEMS based inertial sensors like accelerometers and gyroscopes. These sensors are also entering in the consumer market, particularly in mobile phones. Gyros can give the image stabilization and GPS system. Inertial sensors have reached a market of US$600M in 2004 and will grow more than US$800M in 2008.

Recent trends are also to make electronics on the MEMS chip if the requirement is in very large scale otherwise hybrid systems are made. Due to the involvement of very high cost to set up an industry for MEMS fabrication, the existing IC industries have been upgraded to make these devices. Many processing steps like growth of silicon oxidation, silicon nitride, photolithography, metallisation etc. are common and already exist in these industries. But there are some difficulties for making many MEMS in the existing foundry. Materials like ferroelectrics, piezoelectric and several dielectric materials have also to be used in making these structures. These materials cannot be used in CMOS Foundry and require additional facilities. Secondly, MEMS technology does not have a common technology base. For example, to make an accelerometer and MEMS based night vision system, different technology and materials are required. Due to these difficulties, a standard technology is needed which can be used across different manufacturers. Although many tools are available to design these systems, more efforts are required to develop special tools. Similar difficulties are with the packaging, and characterization facilities for each device are different.

Recently developed nanotechnology has given big hopes to yield many new structures, which will lead to the new revolution. Technology has been developed to make nano-structures by building up atoms and molecules using bottom up approach and the functionality developed in these have been studied from the materials point of view. MEMS technologies can make it possible to achieve the applications of these nanomaterials by combining the functionality of these nanomaterials with micron size devices.

It is now most important to develop technology to join biomaterials with MEMS. These biomaterials have high compatibility with biological functions. These bio-functional devices need not to be characterized by chemical reactions but by mutual physical interactions. This type of integration can also provide a variety of functions.

Due to the possibility of implantation of these micro-sensors inside the body, it is also necessary to have the information to be transmitted by telemetric system. Therefore wireless technology is to be embedded into these bio-MEMS systems. The workshop has also covered this topic in a detailed manner. I feel that it is a great success of the NAM S&T Centre in organizing this workshop and choosing such an important topic of the workshop. Although the presentations are from different nations and many papers contain their country status with technical details on this subject, it will be highly useful for many others as well as will be useful in networking among different countries.

V.K. Jain
Emeritus Scientist
Central Scientific Instrument Organisation
C.S.I.R., New Delhi

Contents

Introduction

Microelectronics is the cornerstone of the information technologies that pervade virtually every aspect of contemporary life. It is difficult to imagine any field of science or technology that has had a more profound impact on the latter half of the Twentieth Century than microelectronics. Microelectronics industry has been able to provide transistors, chips and products that are becoming smaller, faster, cheaper and better every year. As transistors become smaller, they become faster, more and more of such transistors can be packed on a chip, and thus chips are able to store and process more information. This has revolutionized all aspects of our lives from our common household appliances to the most advanced weapon systems.

However, the information technology is not restricted to merely to simply pack more transistors onto a chip. Further technological revolution will involve the incorporation of new structures, including microscopic machines on a chip alongside the transistors creating a whole new generation of computer chips and a chip that cannot think, but sense, act and communicate as well. These fully functioning machines have sizes of the order of a human red blood cell or even smaller. Advances in microelectronics, information processing and communication are bringing in a new era not just in technology but also in all the areas of life: art and culture, human relations and education, economics and lifestyle, perception and cosmology. The personal computer has radically changed how we think about and use information, how we act, and how we interact. These computer and communications technologies are the basis for changes involving automation, energy conservation and pollution control in offices, factories, automobiles and homes; supercomputers for applications from weather prediction to computational research; new capabilities in financial services; new means of storing and playing back audio and video recordings; advanced telephone and television systems; and complex weapons systems for national defense. Each of these areas is critically dependent on microelectronics technology. But even relatively simple products such as the telephone answering machine, VCR and microwave oven have also altered our sense of time, our expectations and our daily lives. More sophisticated products such as cellular phones, compact disks, portable video cameras and digital photography are having and will continue to have a yet more profound impact on what it means to be human.

Recent progress in microelectromechanical systems – the microelectronics, microfabrication and micromachining technologies, collectively known as MEMS – is being applied to biomedical applications and has become a new field of research unto itself, known as Bio-MEMS. The technology is originally based upon the same technology that has been used to make computer chips ever more powerful and less expensive. MEMS technology has enabled low-cost, high-functionality devices in some commonly used areas, such as inexpensive printer cartridges for ink jet printing and chip-based accelerometers responsible for deployment of automotive airbags. Bio-MEMS applies these technologies and concepts to diverse areas in biomedical research and clinical medicine and is an enabling technology for ever-greater functionality and cost reduction in smaller devices for improved medical diagnostics and therapies. The inherent characteristics of Bio-MEMS promise the production of highly miniature, smart and low-cost biomedical devices that could revolutionize biomedical investigation and clinical practice. Consequently, a primary thrust is the research into development of Bio-MEMS and associated nanotechnology for clinical applications such as surgical instruments, tissue repair, artificial organs, diagnostic tools, and drug delivery systems.

Considering the ever increasing importance of Microelectronics in our day-to-day life and the need for capacity building in this area in the developing countries, the Centre for Science and Technology of the Non-Aligned and Other Developing Countries (NAM S&T Centre) identified the subject as one of the priority areas. In this context, the Centre held four international workshops during 2000-2004, respectively on 'Developments in Microelectronics' during August 21-24, 2000 at the Asian Institute of Technology (AIT), Bangkok, Thailand; 'Trends in Microelectronics R&D and Industry' during November 20-23, 2001 in Da Nang, Vietnam; 'Development of Microelectronics: Role of VLSI in MEMS' during December 4-6, 2002 in New Delhi, India; and 'Microelectronics: Role of Developing Countries in VLSI and Micro-Systems Technology' during December 24-26, 2003 in Cairo, Egypt. This series was continued with the organisation of the 5th international workshop on Microelectronics on the theme 'Wireless Technology and MEMS' during Sept 6-7, 2005 at Kuala Lumpur, Malaysia jointly with the Ministry of Science, Technology and Innovation of Malaysia (MOSTI) and MIMOS Berhad. The workshop was aimed at exchanging country specific information on MEMS and wireless technologies; establishing a network for research in these areas across the NAM and other developing countries; facilitating the developing countries to establish relations for joint industrial projects in these areas; and exploring other models for microelectronics industrial development. It covered the areas of Bio-MEMS, 3G wireless technologies, RFID technology, analog mixed signal design and process, and RF technology.

Twenty six scientists and senior experts from 15 countries, viz. Australia, Cuba, Egypt, India, Indonesia, Malaysia, Mauritius, Nigeria, Pakistan, Philippines, Sri Lanka, Tanzania, Turkey, UK and USA attended the workshop either as participants or as resource persons. There were four resource persons in the workshop, namely, Dr. Arif Anwar, Vice President, Synamatix Sdn Bhd from the UK; Prof. Chung-Chiun Liu, Wallace R. Persons Professor of Sensor Technology & Control, Professor of Chemical Engineering and Director, Center for Micro and Nano Processing at the Case Western Reserve University, Cleveland, Ohio, USA; Prof. Sudhanshu Shekhar Jamuar of India and currently with the Department of Electrical and Electronic Engineering of the Universiti Putra Malaysia (UPM), Selangor; Malaysia; and Dr. Tun Zainal Azni Zulkifli, who respectively delivered talks on 'MEMS in Bio Medical Applications', 'Infrastructure Requirement for Bio-Sensing', 'Current Trends of Low Voltage Analog Design for Wireless Applications' and 'Recent Developments and Application of RFID'.

Scientific papers and country status reports were presented by Dr. Luis Miguel Ledo, Specialist, Centre of Applied Technology & Nuclear Development (CEADEN) of Cuba; Dr. Hamed Abdelghafar Hamed Elsimary, Chairman, Microelectronics Dept., Electronics Research Institute, Cairo, Egypt; Dr. Vinod Kumar Jain, Emeritus Scientist, Central Scientific Instrument Centre and Mrs. Sanikommu Nirmala, Scientist, International Advanced Research Centre for Powder Metallurgy & New Materials of India; Dr. Totok Soegandi, Director and Dr. Golb Wiranto, Senior Researcher of the Research Centre for Electronics & Telecommunications, Indonesian Institute of Sciences; Ms. Bharti Kisto, Licensing Officer of the Ministry of IT & Telecommunication in Mauritius; Dr. Ashiru Sani Daura, Director, National Information Technology Development Agency (NITDA) of Nigeria; Mr. Muhammad Malik Asif, Assistant Professor, COMSAT Institute of Information Technology, Pakistan; Dr. John Richard E. Hizon, Assistant Professor, Microelectronics and Microprocessors Laboratory, Department of Electrical and Electronics Engineering, University of the Philippines in Quezon City; Mr. Ananda Sisirakumara Pannila, Head/Technical Manager, Industrial Technology Institute, Sri Lanka; Mr. Emmanuel John Kaale, Scientific Officer (ICT) Grade I in the Tanzania Commission for Science & Technology (COSTECH); and Ms. Gul Senturk, Researcher in the National Research Institute of Electronics and Cryptology of Turkey, Dr. Ahmed Shuja, Associate Professor, COMSAT Institute of Information Technology, Pakistan.

Panel Discussion was held in two sessions, to explore joint research opportunities in Bio-MEMS in the developing countries and for joint development of wireless products. Dr. Waleed Faris of IIU was the moderator of the 5-member first discussion session with Prof. Dr. Burhaniddin Yeop Majlis (UKM), Dr. Othman Sidek (USM), Dr. Arif Anwar (Synamatix) and Prof. Chung-Chiun Liu (CWRU, USA) sharing the dais. For the second discussion session Dr. Ahmed Mabrouk of MIMOS was the moderator and Dr. Tun Zainal Azni Zulkifli (USM), Prof. SS Jamuar (UPM), Dr. Abu Khari A'ain (UTM) and Dr. Sze Wei Lee (MMU) were the members.

Finally, a Resolution with the following recommendations was adopted for implementation in the future:

(a) Focus on commercializable products: R&D push and end-user/application pull must be considered.

(b) Generate database of wireless and MEMS researchers and research areas among NAM members. This can help identify areas of focus/interest, and quantify human resource strength.

(c) Form specific teams, which consist of government, researchers and end-users to oversee collaboration efforts in focus areas and commerciable products.

(d) Leverage existing facilities in NAM member countries for specific technical activities such as fabrication, characterization and packaging. Build additional facilities for specific requirement in appropriate locations/ countries.

On behalf of the Centre, I would like to express our sincere gratitude to Mr. Iskhandar Md. Nasir, Manager (Wafer Fabrication) in MIMOS, and Mr. Muhammad Matori Bin Ali and Dr. Ahmad Wafi Harussani of MOSTI who had been the real workhorse behind this event on the Malaysian side. I also pay my gratitude to our Malaysian partners who received excellent planning support from their senior colleagues, Mr. Mohamad Akbar Mahbat, Director of the International Division in the Ministry of Science, Technology & Innovation of Malaysia, Mr. Wan Mohd. Salleh, Vice-President (Microelectronics and Semiconductors) in MIMOS and other members of the organizing committee of the workshop. Thanks are also due to the MIMOS for their partial financial support to the NAM S&T Centre for this publication.

I would like to acknowledge the dynamic involvement and untiring efforts of Dr. V. K. Jain for technical editing of the book and last but not the least, the untiring efforts of the entire team of the NAM S&T Centre, particularly Mr. M. Bandyopadhyay, Mr. Gaurav Gaur and Mr. Pankaj Buttan which culminated in these proceedings and in compiling the presented papers and giving a shape to the volume.

I hope that the volume would serve as a reference material for the professionals working in the area of microelectronics and will help the developing countries in working out appropriate plans, strategies and exploitation of new technologies and research results for the utilization and prosperity for the masses.

Dr. Arun P. Kulshreshtha
Director, NAM S&T Centre

1

Biomems and their Applications

Vinod Kumar Jain
Emeritus Scientist,
Central Scientific Instrument Organisation (Delhi Centre),
C.S.I.R Complex, Pusa Campus, New Delhi – 110 012
E-mail: *vk_mems@yahoo.com, mems_20032003@yahoo.co.in*

ABSTRACT

Micro electro Mechanical Systems (MEMS) have applications in every field, but major areas are coming like in aerospace, defense, biomedical and communication. BioMEMS are making revolution in biomedical investigations and in clinical use. Present investigations are concentrated in the field of catheters, blood pressure testing and blood analysis. The paper presents some of the advanced applications of BioMEMS based catheters for cardiology and also given details of the investigations made here to find out the hot spot causing vulnerable plaques in coronary. It is also important to monitor the Intra Cranial Pressure (ICP) in neurological patients, a wireless telemetric transfer of data, mems based pressure sensor has been presented. The paper also presents the requirement of the microneedle to collect the very small quantity of blood from the human body without any pain and fear like a mosquito. It is also important for making the lab on a chip. Microcantilever is the latest in the bio-detection system. When a molecule absorbs at its surface, the microcantilever beam bends due to change in the mass and it can be measured optically or to detect the change in the resonance frequency. Modeling has been done to find the mass sensitive factor has been calculated. Calculations shows that 7.5 pgm mass can be detected using the particular dimensions of the microcantilever. These microcantilevers can be made using bulk as well as surface micromachining.

Introduction

The next revolution in silicon technology is presently due to the development of MEMS and it is different from the old, which was only to package more transistors onto the silicon chip. The power of the chips will not be limited simply to store and process the information. Incorporation of new type of functionality on the chip is making the structures to sense, act, and communicate as well.This is making a chip to have ability to have informations from surroundings and communicate what they know. Now it is important that how much chip is capable to understand its environment, affects its environment, and benefits its user. MEMS are intelligent micromachines having sensors, actuators and signal processing electronics at one place. This technology has revolutionized every field by bringing together silicon based microelectronics and micromachining technologies. MEMS is known by different names as smart systems, micro systems etc. these are non-biological systems but analogous to biological systems.Human body is the perfect example of a smart system, having sensors, actuators and brain as intelligent control. Consider the same system at micro scale that a mosquito or an insect having the same systems. A machine developed of the insect size having the same functions called MEMS.

No machines are available to make the very small thicknesses of the membranes or can machine a particular shape to make a micron size device. Bulk and surface micromachining of silicon has made possible to decrease the thickness of the structure to a few microns or even less than a micron and can give the required shape of the device. These machines have the capability to perform a variety of functions like physical and chemical sensing, analyzing the information and communication. The most important factors are that these devices are very small in size and also of low cost as these devices can be batch fabricated in thousands at a time.In these devices inertia and gravity is less important than effects of atomic forces surface science. Now, integrated circuit manufacturing industries are geared up more to make MEMS devices, as many process technologies are same as growth of silicon dioxide, silicon nitride, and alumunium or gold coatings. But for MEMS, fabricating some materials outside this limited materials are required and introduction of other materials like ferroelectric and Piezo electric and several dielectric materials is risky to be use (in the CMOS Foundry).Secondly MEMS technology does not have a common technology base. For example, to make an accelerometer and MEMS based night vision system, different technology and materials are required. In view of this requirement, many standard I.C. Fabs have been upgraded to take up MEMS production. Recent trends are also to make electronics on the MEMS chip if the requirement in on very large scale, otherwise hybrid systems is made. Due this revolution, there is a need to have the standard technologies that can be used across different manufacturers. Although many tools are available to design these systems, but more efforts are required to develop more special tools. Similar difficulties are with the packaging and characterization facilities for each device, it is different.

MEMS devices have applications in every field, but major areas are coming like in aerospace, defense, automobiles, biomedical and communications and in many other sectors. Here we have developed technologies for several devices like IR sensor for night vision, pressure sensors, micro turbine and for finding the hot spots in artery for cardiology. Some other devices are under process for neurology for Intra Cranial Pressure (ICP) measurement. This paper presents the review and some of the work done here in the area of BioMEMS.

MEMS-is being applied to biomedical applications and known as BioMEMS. The basic properties of BioMEMS promise the manufacturing of miniature, smart, intelligent and low cost biomedical devices and these devices are making a revolution in biomedical investigation and clinical use. Some of the devices like blood pressure measurement system, etc. based on this technology has captured a big market. According to a forecast by the Nexus Task Force, the market for BIOMEMS is expected to reach a few billion dollars in 2005. The commercial success of these devices and the technical potential of BioMEMS compel to do more research in this area. Some of the biomedical devices based on this technology are DNA and protein analysis chips, lab on a chip, drug delivery systems, miniaturized sensors for catheters, micro cantilevers for chemical and biological sensors, microneedles and thermal imaging, micro cutting tools etc. The paper presents some of the recent developments in this field.

Basic Technologies

Apart from the silicon IC fabrication facilities like diffusion, oxidation, metallisation and photolithography, technology of micromachining of silicon is required. Micromachining can be classified in two major categories *i.e.* bulk micromachining and surface micromachining. Bulk micromachining which is isotropic and anisotropic types can be used to make beams, thin diaphragms, cantilevers etc. Surface micromachining is generally used to make freely suspended structures on the silicon chip. Some times both the technologies have to be used to make many complicated structures. Deep reactive ion etching system (DRIE) is also used in case of high aspect ratio is required. Although thin membranes, beams, cantilevers have been made here to make pressure sensor, vibration sensors and accelerometer, but following two typical devices developed here, show the utilization of bulk and surface micro-machining.

Micro Air Turbine Using Bulk Micromachining (Figure 1.1)

A new type of microflow meter based on MEMS technology has been made here by Jain *et al.* The 60 μm thick silicon rotor has six blades and can rotate with a speed of 50,000 rpm with the help of air jet. The system has been housed in a silicon cavity of about 200 μm deep in the silicon wafer. The silicon cavity has two fine channels one for inlet and other for outlet of the gas. Complete modeling has been done for proper angles of the inlet and outlet. An optical photovoltaic sensor of 200 mm radius has been generated in the cavity of silicon under the blades of the turbine. The number of the output pulses of the sensor due to the chopping of the incident light, by rotation of the rotor, has been correlated with input flow of the gas. The control circuit commands the input flow based on closed loop system. This micro flowmeter (Figure 1.1) has the response in microsecond. It can be used in many industrial applications and also in medical uses.

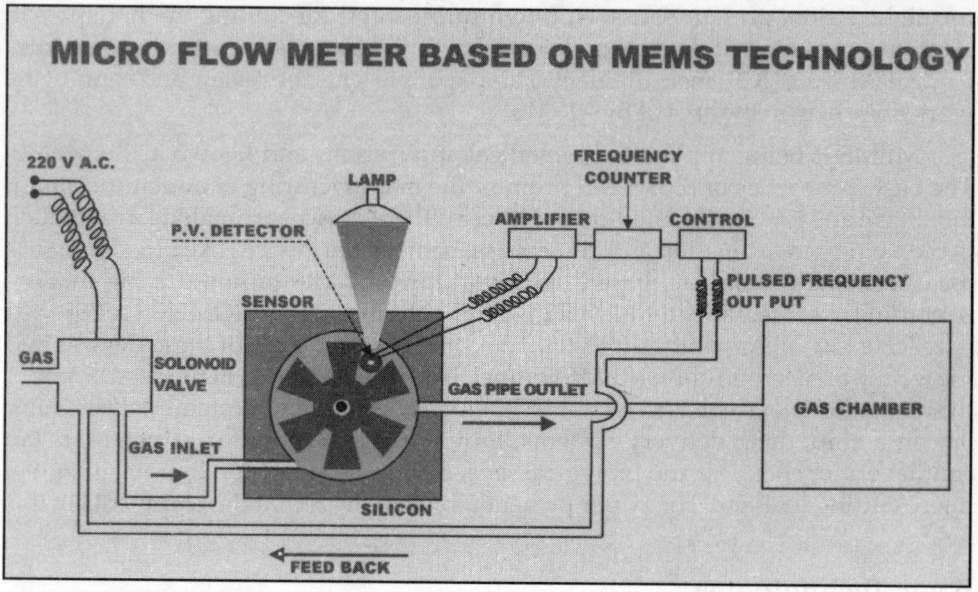

Figure 1.1

MEMS Based Infra-red Sensor Array Using Surface Micromachining

Uncooled infrared (IR) sensors are entering in a new phase due to their large number of applications both in military and many commercial sectors. Apart from their uses in night vision system these are thought to be used in micro-vehicles, robotics, and in biomedical for thermal imaging. These detectors are gaining importance over the HgCdTe photon based detectors because of their room temperature operation, low power consumption and smaller in size. Surface micromachining has made it possible to make these IR sensors. We have developed these sensors using titanium bolometer as the IR sensing elements on the freely suspended silicon nitride membranes (1.0 μm thick and 50 × 50 μm in area) on the silicon substrate. The membrane is supported by two small hinges at the corners and has air gap of 2 μm between membrane and the silicon substrate. This provide a thermal isolation which is a very important requirement for making a IR sensor. It has been made using surface micromachining. The technology is quite difficult, but here it has been perfected and transferred to a industry to make the array having a large number of elements. Finally the sensors are to be integrated with CMOS on the same chip. The array of 16x16 made is shown in Figure 1.2. The more detail, of design and fabrication process can be seen from our previous publications.

Bio MEMS

Micro Needles

Blood test is most common in case of health monitoring. It is quite painful and fearful when skin is pricked by a needle to take the blood for test. In 2001 Oka *et al.*,

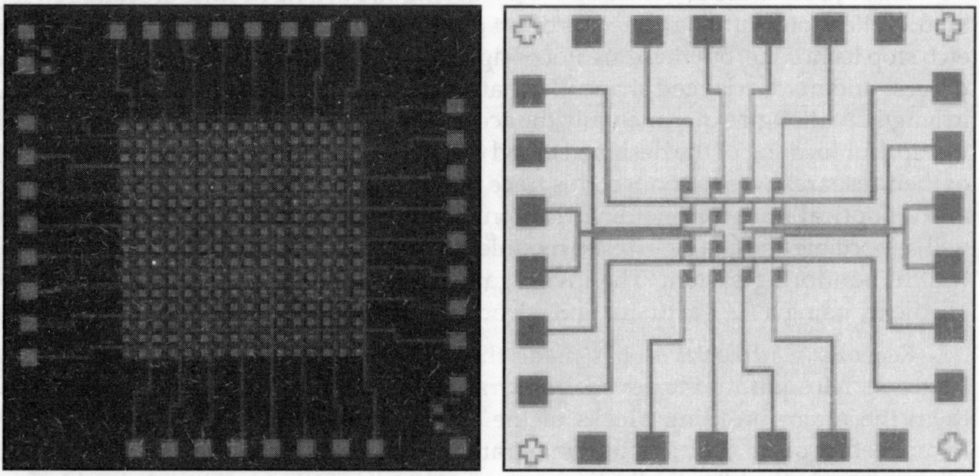

4 × 4 Bolometer IR Array **16 × 16 Bolometer IR Array**

Figure 1.2

thought that the pain and fear could be reduced, if a micro needle can be developed for the blood collection. Again, inspiration came from Nature seeing that one do not feel pain and even do not know when a mosquito sucks blood from the human body. According to them, the mosquito needle has a jagged shape and it can easily cut or penetrate in to the skin with the help of vibrations. With this idea Oka *et al.* made efforts to fabricate a micro needle shown in Figure 1.3(a):

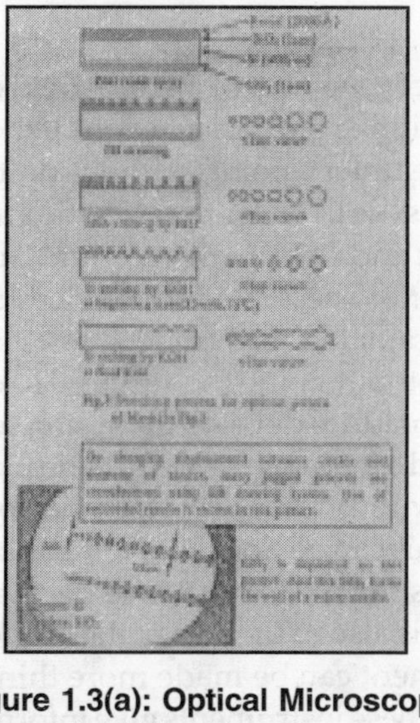

Figure 1.3(a): Optical Microscope Photograph of Negative Groove, which is Obtained by Searching Process

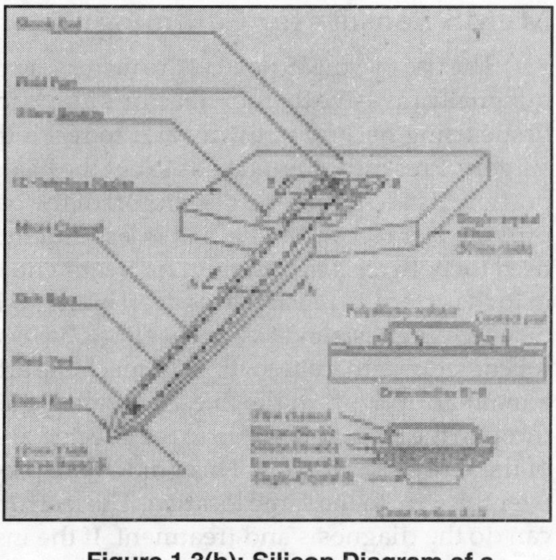

Figure 1.3(b): Silicon Diagram of a Silicon Processed Micro Needle

These needles are of generally 200 μm in length, but long needles are required for blood collection. Although needles of 2-4 μm in length have been made using boron etch stop technique, but these are not of jagged type. Oka *et al.* used a new fabrication process and made a jagged shape like that of a mosquito of silicon dioxide and 1 μm in length. Although experimentally the accurate jagged shape was not achieved near the tip, but forming of the designed blood collection system was succeeded. The goal of their research was to connect this trace of blood collection system to a blood tank and an optical fibre for photometric monitoring of blood components. The system will be portable and can be wireless module. It was expected to use this as a continuous health monitoring system. The investigators were able to test the strength of the needle by using a PZT actuator and a load cell.

Recently Gattiker *et al.* has made the similar needle and given the name as Electronic Mosquito as the device can extract the blood like a mosquito. According to them the major building blocks of the MEMS based proposed system will be microneedle, piezoelectric acuter integrated with the microneedle, electrochemical microsensor in the blood collecting compartment, control electronics, microbattery and an antiseptic layer placed between the skin surface and the device. CMC expect with the team to make a MEMS based system, which will regulate insulin levels automatically for the people having diabetes.

Figure 1.3(b) shows the microhypodermic needles that have been made using bulk and surface micromachining by Lin & Pisano. These microneedles have also micro pump and electronic circuits on board for drug delivery and can also collect the very small quantity of fluid for testing. The channels are 9 μm in height and have micro holes. These microneedles have been tested in animals without damaging any muscle tissue. Microelectrodes can also be made using the similar technology required for the study of brain.

MEMS Sensors for Cardiology

The use of MEMS devices are increasing in a very big way in medical field due to minimally invasive therapy. Doctor's aim is to minimize the damage done to healthy tissue using natural pathway and to reach the treatment point, without doing the surgery. Presently surgeon has to cut the healthy tissue in order to reach the affected portion. It can reduce the discomfort of the patient and risk of infection and can also provide fast recovery. The cost is less and more safe. These types of devices can be used for both for diagnosis and for treatment. Like catheters and guide wires, inserted in to the artery, to reach places deep in the body, without doing the surgery. To do the detailed diagnosis and treatment efforts have been made to develop the silicon sensors for catheters and guide wires. To reach the affected area, an incision is to be made in a main artery. A thin flexible wire called guide wire is to be inserted and to push through the blood vessel, up to the affected area. This can be done under the guidance of the x-ray fluoroscopy. This guide wire provides the guidance to the instruments over the wire to the same location. The instrument is also of the size of the guide wire can do the diagnosis and treatment. If the instrument can be made more thinner, it can be passed in to the more smaller blood vessels. These instruments give information about the positioning of the catheter and the guide wire, diagnosis, monitoring of the treatment and checking of results of the treatment. It is required urgently to develop a large number of many other sensors small enough to put at end of the guide wire or on catheter tip. These sensors will be able to give information like cross section of the blood vessel, blood pressure, flow rate, chemical analysis such as oxygen level in blood, temperature of the blood and many other informations. The multi –sensor chip approach can provide much more information and it can be thought to fabricate with monolithic approach. Limitation in fabrication of the monolithic chip may be due to same technology should be used to make all sensors.

Jin Ji *et al.* developed a multiplexed ultraminiature pressure sensor for a cardiovascular catheter. Pressure sensors have many applications in health care and especially to find out pressure and flow of blood in the cardiovascular system. Worldwide angiography or catheterization is the major diagnostic tool for finding the blockade in the artery. The sensors are typically at the external end and measuring pressure through long fluid-filled lumen may restrict very accurate measurement. Although catheters are available with pressure sensor at the tip, but at a very high cost. Jin Ji *et al.* proposed that measurement of the pressure gradient in the cornary arteries of the heart. According to them, these pressure sensors should have high resolution of the pressure and should be extremely small in size. MEMS technology has made it possible to fabricate these sensors.

Figure 1.4 shows the structure of the pressure-sensing catheter, which has only two lead wires. These capacitive pressure sensors have been made using silicon micromachining technique and mounted in the wall of the catheter as shown in the figure. The Figure 1.4 also shows the pressure sensing capacitor is formed between the metal pad on the glass substrate and the diaphragm made in silicon. The entire structure is 350 μm wide, 1.4 μm long, and 100 μm thick. CMOS readout electronics there converts the sensed pressure in to a frequency variation signal and the device can resolve the pressure variation of about 3 μmHg.,while vacuum sealed devices should permit <2 μmHg.

In another applications of bioMEMS in cardiology, it has been observed that during the angioplasty surgery, the balloon catheter is used in the operation theatre. Doctors can determine the balloon pressure only from the out side pressure gauge or injector's gauge. There is no other technique to find out the absolute pressure inside the balloon when it is dilated. It can be very risky, if the actual pressure in balloon inside the artery is not known. There will be some difference in the pressure as the balloon is in the blood vessel where blood is flowing and have its own pressure on the balloon. Tohyama *et al.* developed a MEMS based pressure sensor with fiber optics and applied them to measure the actual pressure in side the balloon catheter. The silicon microsensor of the size of 270 × 270 × 150 μm has been made using silicon micromachining and anisotropic detecting. The sensors have to be used inside the human body, it is required to have a more simplified measurement system. High order of detection can be obtained using the fiber interferometers. According to them, they require more complex techniques, while intensity modulated fibre-optic sensors require only a simple measurement system. For the transmitting source and signal light, a multimode optical fibre is used as shown in Figure 1.5.

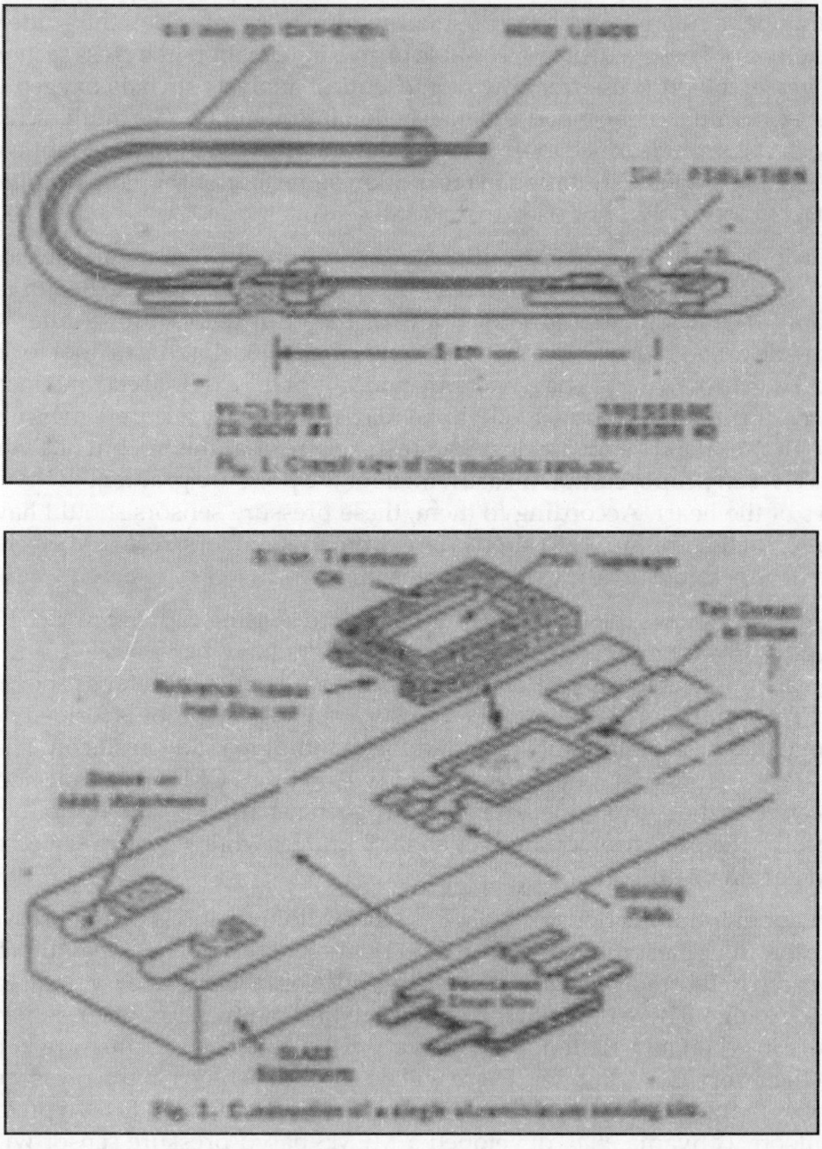

Figure 1.4: Two Site Pressure Sensing Catheter Pressure Sensor Based on Capacitance Change

Figure 1.5 shows the balloon catheter with the fibre-optic pressure sensor. They have injected the pressure in the range of 0.1-0.5Mpa to the balloon, and the pressure at the outlet of the pump and inside the balloon was measured simultaneously. They found a large difference in pressure inside the balloon and the injector. Their finding is highly useful to find out the exact pressure inside the balloon while dilated.

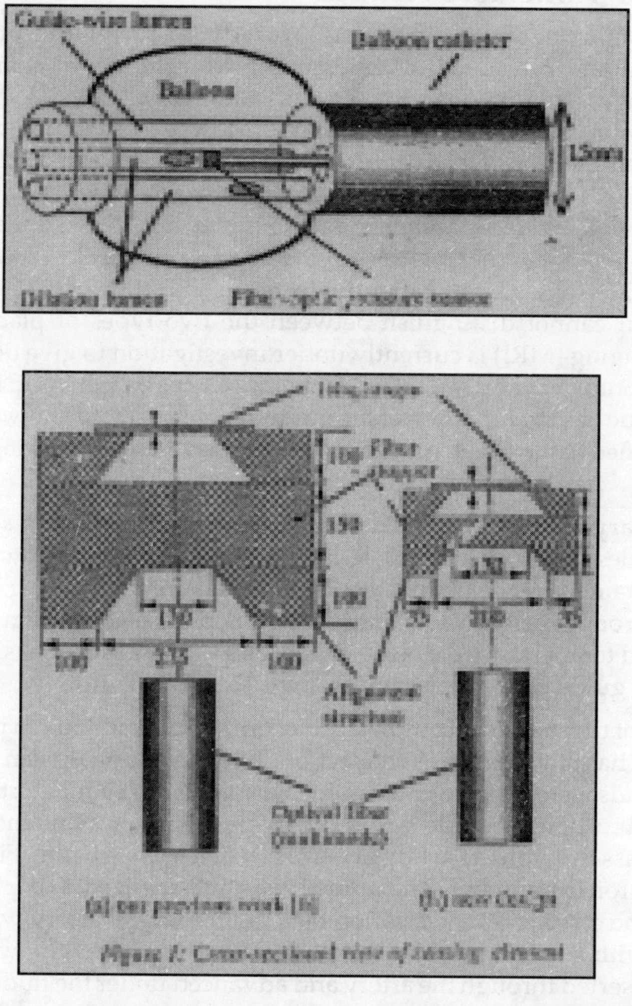

Figure 1.5: Sensing Element Balloon Catheter with MEMS Sensor

Coronary Thermography

At present, the atherosclerotic plaque is considered as a mass in the coronary artery, which obstructs the lumen of the artery. The plaque can contain active cells, which play an important role in its stability. According to C. Stefanandis *et al.* of university of Athens, Greece, the stable plaque have thick fibrous caps while vulnerable plaques have thin fibrous caps, which have high risk of rupture. Due to the rupture of these atherosclerotic plaques formed, the clots causing the heart attacks, but according to them still these happenings cannot be reliably predicted. The major problem is the detection of these plaques, which are vulnerable and have high risk.

Detection of Vulnerable Plaques

The aim is to provide diagnostic procedures to give information about morphological and physiological assessment. Morphological characteristics are related to changes in the coronary vessel walls like area etc. Physiological assessment gives coronary pressure, flow, temperature etc. Most of these techniques are invasive and there is no method, which can provide all these information. Intravascular ultrasound (IVUS) is a catheter – based technique that provides tomographic images of the arterial wall and morphometric data of lumen and vessel dimensions. Recent studies have shown that the thickness of the fibrous cap may be visualized by high-frequency IVUS. Angiography can provide the visualization of the plaque with high sensitivity, but cannot distinguish between the two types of plaques. Magnetic resonance imaging (MRI) is currently under investigation to give details about the plaque in a noninvasive way. Optical Coherence tomography (OCT) is a catheter-based technique having reflection of the infrared light in situ to know the information about the calcified tissue. Some preliminary results have come regarding the vulnerable plaque.

Preliminary studies using Thermography equipment has shown that the coronary plaques, which are vulnerable, have higher surface temperature. Some results have been obtained by Stefanadis during his measurements using Thermography and given a strong possibility to identify the vulnerable plaques. Jain *et al.* developed a catheter- end temperature probe when the temperature sensor has been bonded at the end of the guide wire at a particular angle shown in Figure 1.6.

The size of the sample is completely biocompatible and 450 μm in diameter. The whole system has only two leads connection. Two models have been tried, in one the connecting leads were taken over the guide wire and in other model; these wires have taken through a hollow guide wire. The over all size was maintained which is required to pass through the artery or coronary of a human being. First, a pig model was selected for experiment. The animal was anesthetized and placed in supine position on the cardiac catheterization table. Left femoral artery was exposed and cannulated with a seven French arterial sheath. The guide wire with the thermal sensor was inserted through the artery and advanced under the fluoroscopy and see that the tip of the guiding catheter touched the wall. Temperature at the site was recorded. The whole assembly of the guide wire and sensor was gradually withdrawn with measurement of temperature of descending thoracic and abdominal aorta at regular intervals. The temperature of coronary arterial wall also recorded at the site. In conclusion, by using both types of thermal probes could be safely introduced into the aorta as well as coronary artery through standard coronary guiding catheters and angioplasty equipment. Temperatures can be recorded from various sites. In future, these systems will be packaged as proposed by other workers and more experimental trials will be made.

J. Sorrell and Bhattacharya at university of Colorado proposed that it can be made possible to put more sensors on the wires that can measure local fluid properties, such as pressure, pressure gradient, temperature, and species of concentration levels. They have also proposed a suitable design for an intelligent guide wire. In their

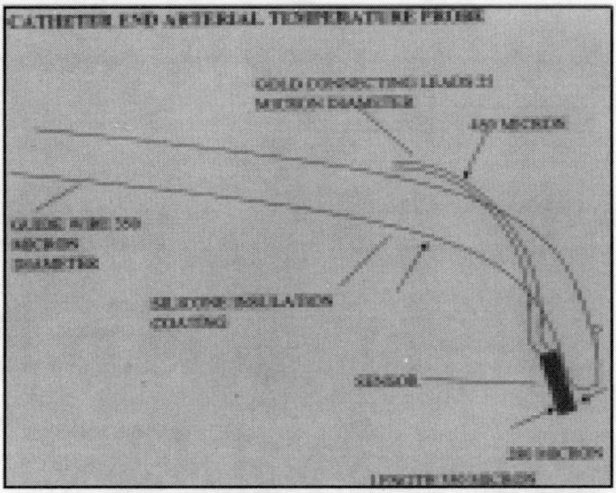

(a) Catheter-end Arterial Temperature Probe

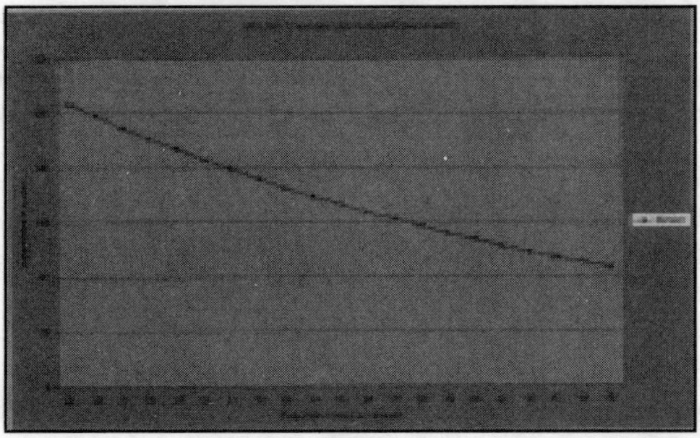

(b) Resistance Variation with Temperature

Figure 1.6

proposed design, care has been taken to the outside diameter should not exceed the present one and should be biocompatible. Four leads should be sufficient to assess many sensors and these connections should also be robust to ensure reliability and safe functioning of the device. The design is shown in Figure 1.7.

It consists a hollow guide wire and a silicon sensor. The sensor can be made using the standard silicon micromachining. The other silicon substrate spacer acts as an electrical bride between the MEMS sensor and the lead wires. Wet etching was used to make cavities in the substrate and these cavities provide the soldering locations for the lead wires and the sensor. The wire housing proposed in the design consists of four lead wires and can be housed in an insulating polymer material. The assembly

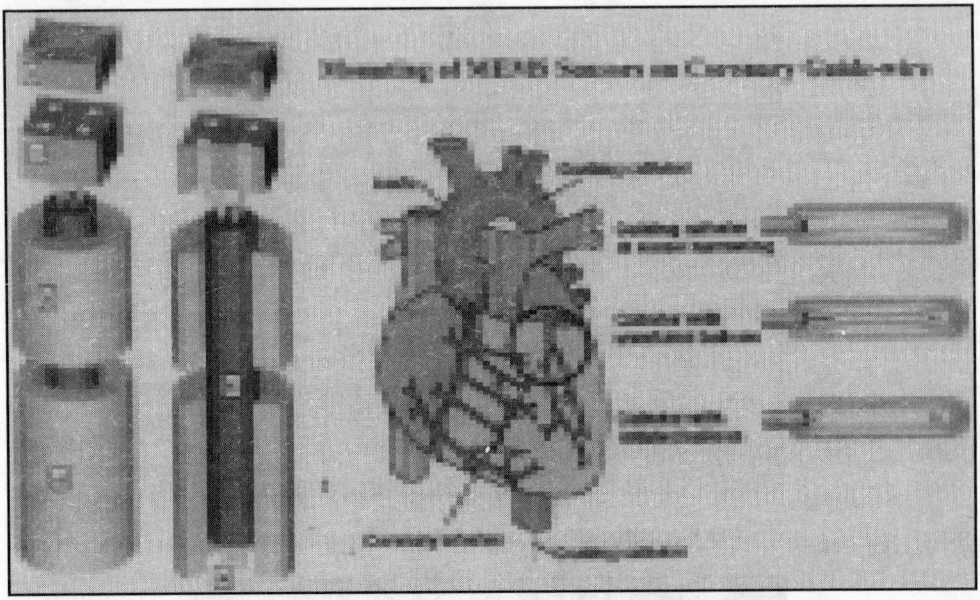

Figure 1.7: Proposed Design for the Temperature Probe and also Other Sensors at the Tip of the Catheter having Connecting Leads through a Hollow Guide Wire

can be run through the hollow guide wire. The lead wires are positioned in their respective cavities of the carrier. The other coating can provide biological compatibility as well as smooth surface that allow the guide wire to move easily through the arterial system.

Neurology

MEMS Based Implantable Intra Cranial Pressure Monitoring (ICP) System

All neurosurgical patients do require ICP monitoring and can avert fatal emergent situations. Continuous monitoring of Intra Cranial Pressure (ICP) with implanted sensor is essential for understanding of many physiological and pathological function of brain. Methods for measurement and continuous monitoring of ICP have been in use for more than two decades. Catheter tip transducer system and strain gauge sensor are currently in use at many places. Each of these systems has their merits and demerits. Further the chances of infection in conventional catheter type sensor are significant higher. An optimal device would meet various specific requirements, of sensitivity, tolerance, drift, hyteresis, pressure and linearity. In addition, it should be possible online monitor without any constraints.

Efforts are here to develop a MEMS based implantable silicon telemetric sensor with a trans-receiver on a silicon chip with the objective to the physiological and pathological studies and to monitor the ICP pressure of the brain sensor can be placed epidurally (Figure 1.8a) and data would be transmitted and recorded on a data

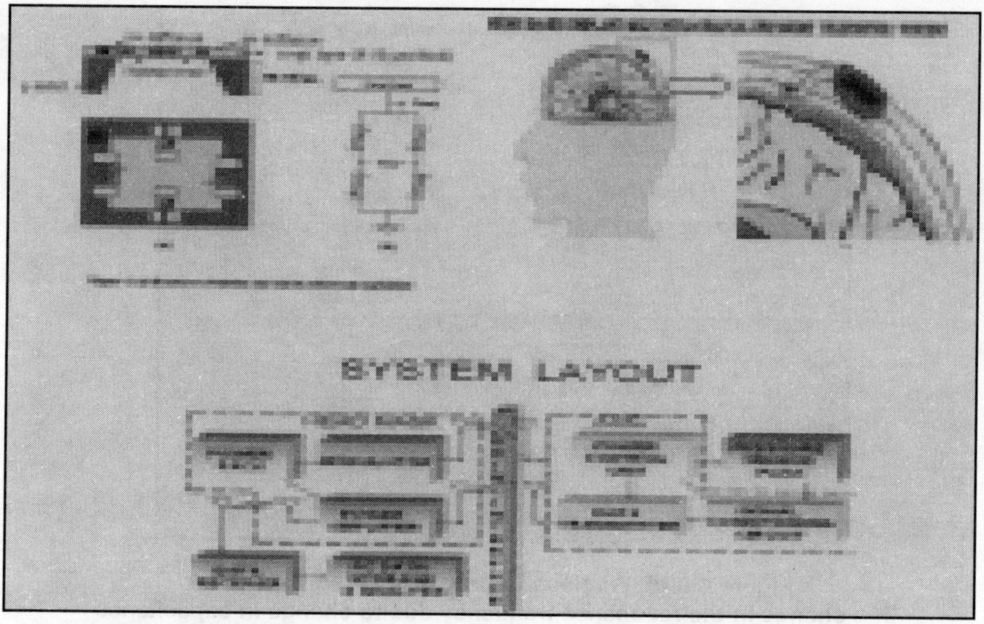

Figure 1.8(a): Telemetric Intracranial Pressure Measurement (ICP) (Ref. 16)

acquisition system. Care should to be taken that IC chip is biocompatible and should be of the required dimensions. Piezo resistive based pressure sensors are the very-first products of MEMS technology. These products are widely used in biomedical applications. A large number of pressure sensors based on this technology have been made here for many other applications. Work is under progress after designing to fabricate the proper dimensions and thickness of the diaphragm using bulk micromachining to measure the ICP pressure. Diffusion, oxidation, metallization and photolithography have been used to develop the Piezo based pressure sensors. O. Akar *et al.* has suggested another approach to develop the absolute wireless capacitive pressure sensor. They have used a parallel inductor-capacitor (LC) resonant circuit for measurement and communication of the pressure. Figure 1.8(b) shows the structure of the wireless pressure sensor.

Using the etch stop technique and bulk micromachining a thin diaphragm was made to measure the change in pressure using the change in capacitive gap. The plainer coil is also placed on the glass using the gold electroplating. The metal coil was made around the metal capacitor on the glass. The variable capacitor and the electroplated coil inside it form an L.C. circuit. The resonance frequency changes with the changes in pressure. Change in resonant frequency is sensed remotely by using inductive coupling, and it does not require the wire connections to communicate the data and for measurement of pressure.

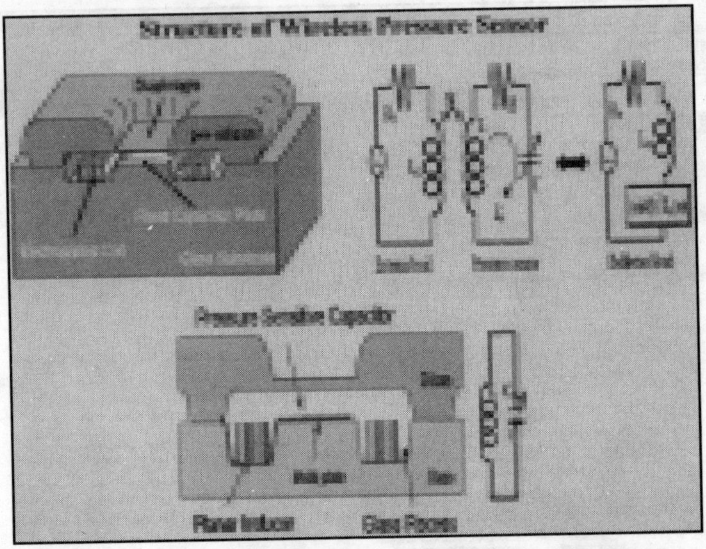

Figure 1.8(b): Wireless Capacitive Pressure Sensor. The change in the resonance frequency due to change in capacitance is sensed remotely with inductive coupling, eliminating the need for wire connections and power supply.

Microcantilever: As a Bio-diagnostic Tool

Microcantilever (MC) sensor technology is a next generation electrochemical technique with a large number of applications in chemical, physical and biological detection. Microcantilever based biosensors can offer many advantages over other biosensors. These microcantilevers (MCs) can be made using bulk or surface micromachining of silicon, silicon nitride or silicon dioxide. In addition to this MCs have an advantage of making of multi-element sensor arrays and can also be integrated with electronic circuitry on the same chip.

The basic principle of operation of a MC is quit simple. When a molecule adsorb at its surface, cantilever beam bends due to the induced stress shown in Figure (1.9). The stress is produced due to change in temperature or may be change in the mass. The bending of the cantilever will be in nanometer scale and can be measured either as the shift in the resonance frequency of the cantilever or by optical deflection method. In the optical detection method, the light beam from a low power laser diode is focused on the free end of the cantilever and its displacement can be known by measuring the reflected light spot with a position sensitive detector. It has certain limitations *i.e.* poor reflectivity of light, long term drift of the cantilever bending and reproducibility in the binding of the analyte to the cantilever surface. Although some of the difficulties have been over come.

Rasmussen *et al.* made the MC with piezoresistive readout. The resistors and electrical wiring were encapsulated using silicon nitride and wiring is done through titanium silicide. Piezoelectric thin films can be easily applied to produce the resonant biosensor in which the added mass of the target bio molecules changes the resonant frequency of the cantilever. The sensor can be produced on the micrometer scale by using lithographic techniques. MCs can also be operated in vibration mode. By loading the cantilevers with extra mass decreases the resonant frequency.

The typical dimensions of commercially available micro machined cantilevers are 50-200 μm long, 10-40 μm width and 0.5 μm-5 μm thick for the mass in the range of a few nanograms. The resonant frequencies of these cantilevers are in the range of a few kilo Hertz to a few hundred kilo Hertz. Depending upon the dimensions the cantilever spring constant varies from 0.008 N/m to 30 N/m. The cantilevers used for AFM purpose are having dimensions of 90-460 μm long, 35-60 μm width, 0.7 μm-7.5 μm thickness, force constant 0.01N/m-91 N/m and 7-420 KHz resonance frequency.

Our present work is an effort to optimize mass sensitivity factor for polymer coated silicon MCs for chemical sensing applications. Simulations are carried for mass sensitivity factor and spring constant change before and after chemical reaction in these cantilevers. The results give the guidelines for designing high sensitive micro cantilevers for chemical and biological sensing applications. For example, in a biosensor, the enzyme functionalized micro cantilever undergoes bending due to change in surface stress induced by the reaction between glucose in solution and GOx immobilized on the cantilever surface.

Analysis

The basic mechanical parameter of a cantilever is the spring constant and the resonance frequency. Soft cantilevers will simply bend due to its own weight and mechanical noise from the surroundings. From the Hooks Law;

$$F = -kZ \qquad \text{............(1)}$$

where,

F-Applied force, k-spring constant, Z-bending of a cantilever. Spring constant for a rectangular cantilever of length L is expressed as;

$$k = \frac{3EI}{L^3} \qquad \text{........... (2)}$$

where,

E-Young's modulus, I-Moment of inertia and L-Length of the cantilever Moment of inertial of simple rectangular cantilever section is given by;

$$I = \frac{WT^3}{12} \qquad \text{...........(3)}$$

where,

W-Width of a cantilever, T-Thickness of a cantilever, I-Moment of inertia of rectangular section and E is Young's modulus. This expression can be used in Eq. (2) to find the spring constant of a simple cantilever

$$k = \frac{EWT^3}{4L^3} \qquad .... (4)$$

The resonance frequency of a long and thin rectangular cantilever beam is

$$f_{res} = \frac{k}{2\pi}\left(\frac{EI}{\lambda}\right)^{1/2} \qquad .... (5)$$

where,

λ Linear Mass density.

Consider a microcantilever of length L_c, width Wc, thickness T_{cs}. The spring constant of the cantilever can be calculated as follows;

$$k = \frac{E_s T_{cs}^3 W_c}{4L_c^3} \qquad (L_c >>>> T_{cs}) \qquad .... (6)$$

where,

E_s-Young's modulus of silicon material

The resonance frequency of the cantilever is

$$f_{res} = \frac{k}{4\pi}\left(\frac{T_{cs}^2 E_s}{3\rho_s}\right)^{1/2} \qquad .... (7)$$

where,

ρ_s Density of silicon material

Chemical Detection Based on Resonance Frequency shift

In this section we present results from theoretical analysis where we utilized the resonance frequency shift of cantilever beam for chemical sensing. In this analysis, we have taken silicon as a substrate material and functional layer as polyvinyl alcohol (PVA). The functional layer is coated on the top of substrate and whose thickness should be less than the thickness of substrate. The reverse case is also considered for analysis in the later part of this section.

After surface modifications, the mechanical characteristics of microcantilever beam changes due to the formation of functional layer. The linear mass density of the modified cantilever now will become:

$$\lambda_f = \rho_s A_s + \rho_f A_f \qquad .... (8)$$

where,

l_f-Linear mass density, P_s-Density of substrate material, ρ_s Density of functional layer material, 'A_s'-Cross section area of substrate material and A_f-Cross section area of functional layer and its flexural rigidity will be redefined as $EI = E_s I_s + E_f I_f$ where s and f represents substrate and functional layers respectively. Now the moment of inertia of the modified beam will also change to:

$$I_s = W_c T_{cs}^3 + W_c T_{cs} \left(Y_c - \frac{T_{cs}}{2} \right)^2 \quad \text{.......... (9)}$$

$$I_f = \frac{W_c T_{cf}^3}{12} + W_c T_{cf} \left(\frac{T_{cf}}{2} - T_{cs} + Y_c \right)^2 \quad \text{.......... (10)}$$

where,

Y_c-Centre of mass, I_s-Moment of inertia of substrate, I_f-Moment of inertia of functional layer, T_{cs}-Thickness of substrate material and T_{cf}-functional layer. The new resonance frequency after surface modification will be expressed as

$$f_{new} = \frac{k}{2\pi} \left(\frac{E_s I_s + E_f I_f}{\lambda_f} \right)^{1/2} \quad \text{.......... (11)}$$

This expression shows that both the stiffness and mass loading affect the resonance frequency. If the functionalized cantilever is exposed to vapor sample, results in swelling of the polymer layer coated on the top of cantilever surface. The mass increase implies decreases in resonance frequency since it is inversely proportional to applied mass. Change in surface stress results in a bending of cantilever. The amount of mass increased after reaction is directly related with the number of molecules deposited or absorbed by the functional layer. Now the final resonance frequency due to the increase of mass cab be expressed as

$$f_{final} = \left(\frac{E_s I_s + E_f I_f}{0.2235 \left(\rho_s A_s + \rho_f A_f \right) L + M L^3} \right)^{1/2} \quad \text{.......... (12)}$$

where,

P_s, P_f-densities of substrate and functional layer. A_s, A_f- cross sectional areas of substrate and functional layer. M-additional mass added at the end of a cantilever beam due to chemical reaction. If we know the order of applied mass during reaction then we can estimate the shift in resonance frequency form Eq. (11) & (12)

$$\Delta_f = f_{new} - f_{final} \quad \text{.......... (13)}$$

Cantilever Fabrication

Materials used to fabricate micro machined cantilevers are silicon, silicon nitride, silicon carbide and silicon dioxide. Composite cantilevers are very much prone to

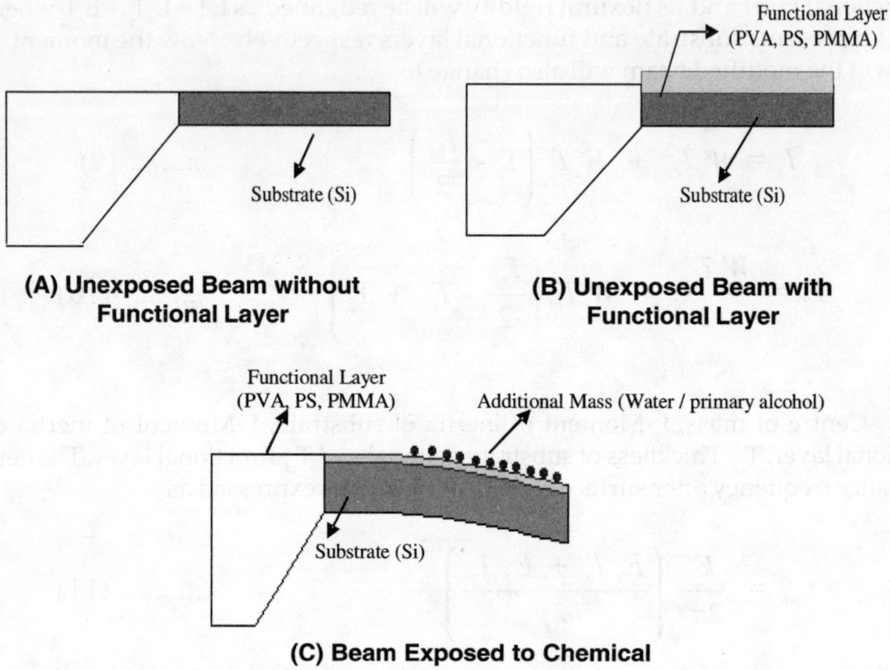

Figure 1.9: Schematic of Microcantilever Based Chemical Sensor

thermal noise due to biomaterial effects. This can be eliminated by optimizing the ratio of substrate to functional layer thickness. The fabrication procedure of simple silicon nitride cantilever is shown in Figure 1.10. In the first step, a small thin layer of silicon nitride is deposited by low pressure chemical vapor deposition (LPCVD) in to a double side polished silicon crystal wafer. In the second step, a deep etch pattern is prepared in the silicon nitride layer on the non cantilever side of the wafer using photolithography and reactive ion etching. Next step is etching of anisotropic surface with KOH until silicon wafer attains to required thickness. Now the required shape is achieved in the silicon nitride surface by reactive ion etching. Finally the cantilever is realized by etching the remaining portion of silicon in KOH and by removing the plasma enhanced chemical vapor deposited (PECVD) silicon nitride with buffer solution. There are many other techniques to make the microcantilevers. The v-shape cantilevers are also under study. To see the mass sensitivity factor, an example of water vapor detection has been taken. The similar results can be obtained by choosing the different functional layers required for biomedical applications.

Detection of Water Vapor

Detection of water vapor plays a vital role in many industrial and metrological applications. To optimise the microcantilever for water vapor detection, cantilever should be coated with water sensitive polymer like polyvinyl alcohol (PVA), carboxy methyl cellulose (CMC) and polyvinyl pyridine (PVP). PVA coated sensor shows larger response to water vapor (Frequency shift up to 120Hz) Typical cantilever

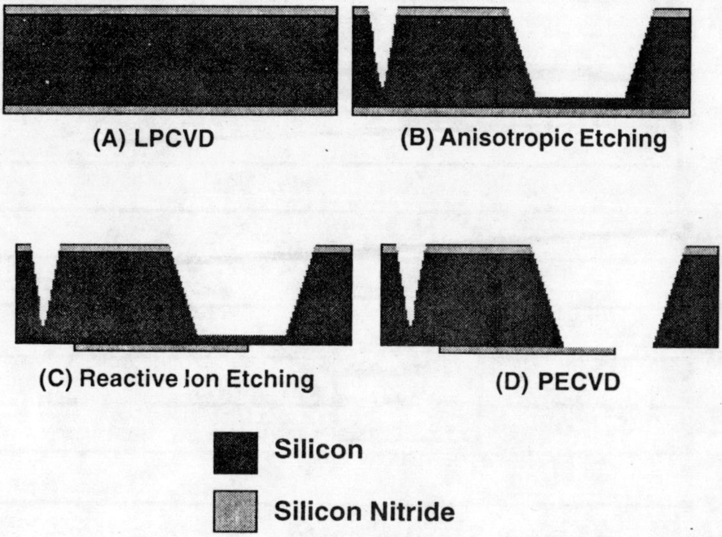

Figure 1.10: Schematic of Silicon Nitride Cantilever

resonance frequency is calculated using equation (10-13) and the same are used for calculating mass sensitivity factor. The mass sensitivity factor is a ratio between shift in the resonance frequency of the cantilevers loading per nanogram. Figure 1.11 shows the graph plotted between mass sensitivity factor and applied mass on the cantilever of different thickness. PVA layer of 2.5 μm thickness on the cantilever has been used as the functional layer to adsorb the water vapors. The result show that sensitivity is almost stable for a particular range of applied mass from 0.41 nanogram to 1 nanogram for a cantilever of dimensions of 500 μm length, 100 μm width and 1 μm thickness. For all other values, the sensitivity increases for applied mass less than 0.41 nanograms or increase if the applied mass is more than 1 ngm. Calculations also show the minimum, which this cantilever can sense is 7.5 pgm.

It also shows that maximum shift in frequency per ngm is 55.92 Hz. The stable range of sensitivity increases with increases in micro cantilever thickness up to 7 μm afterwards it starts to decrease but sensitivity goes down and also insensitive to low mass changes. Hence if the change in mass due to the absorption of water molecules on the top of micro cantilever is in the order of 0.41 ngm to 1 ngm then 500 μm × 100 μm × 1 μm will be the most suitable dimensions for designing the sensor. Similarly, a large number of calculations has been done to optimize the thickness and keeping the length and width as constant parameter.

Figure 1.12 shows the graph plotted between the mass sensitivity factor verses thickness of the micro cantilever and keeping the fixed thickness 2.5 μm of the PVA functional layer. Among the results obtained, important finding is that if the functional layer thickness is greater than substrate thickness, the mass sensitivity is negative, starts to decrease with increase in cantilever mass. This inverse effect can be seen in

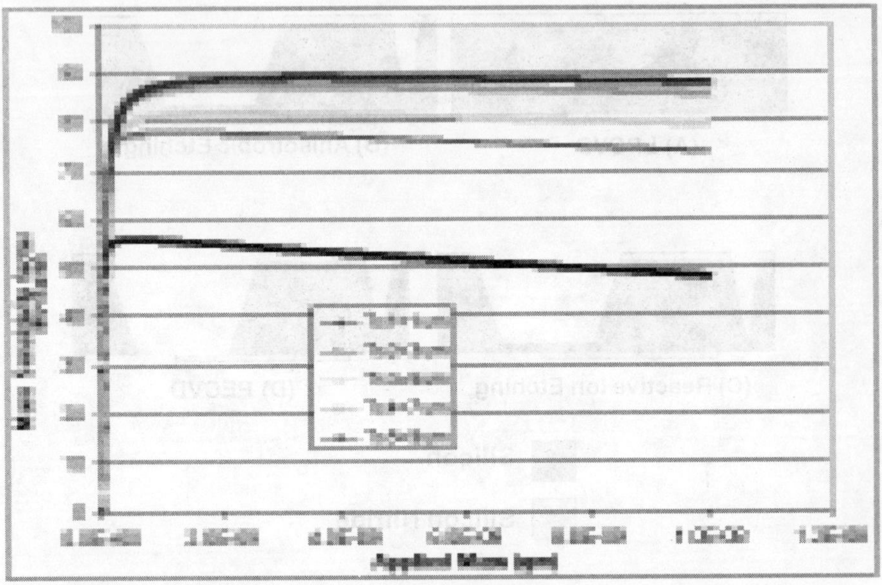

Figure 1.11: Variation of Mass Sensitivity Factor with Change in Applied Mass at Different Values of Microcantilever Substrate Layer Thickness (PVA-500 × 100 × 0.1 µm to 10 µm, Functional Layer Thickness is 2.5 µm)

Figure 1.12 and this is due to the large difference between the substrate and a functional layer young's modulus. This inverse effect starts changes when the substrate thickness approaches to the thickness of the functional layer. These observations suggest keeping the functional layer always less thick than the thickness of the cantilever. Optimisation has been done for the measurement of minimum mass of water using polyphinyl alcohol as the functional layer on silicon-based microcantilevers for chemical sensing applications. It gives the critical observation towards the selection of functional and substrate layer thickness of composite microcantilever.

Conclusion

BioMEMS are making a revolution in the field of biomedical investigations and in clinical use. Micromachining of Silicon has made it possible to make devices of micron sizes, which can be batch, fabricated at a very low cost. These devices have also good reliability. The recent development of the catheter end arterial temperature probe can detect the most vulnerable plaques in the artery. These sensors can be mounted at the catheter end and can be packaged in a hollow guide wire with complete biocompatibility. For ICP monitoring the MEMS based pressure sensor can be made with wireless connections. Micro cantilevers are also under progress, which can detect even a very small number of molecules. The present paper explained some of the sensors for biomedical based on MEMS technology, but there many others like silicon box for diabetic control, micro channels for fluidic control, electronic nose,

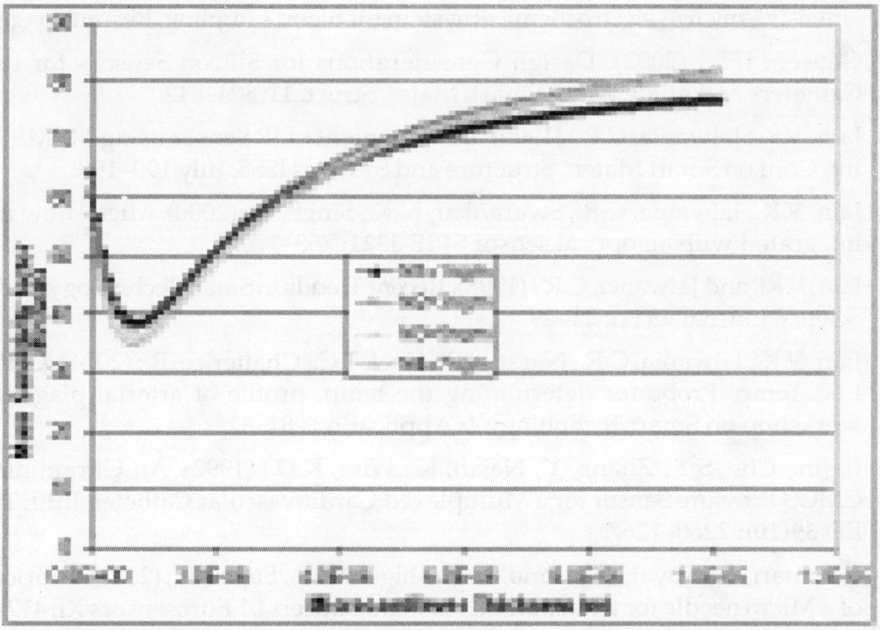

Figure 1.12: Variation of Mass Sensitivity Factor with Change in Substrate Layer Thickness of Microcantilever at Different Values of Applied Mass (PVA-500 × 100 × 0.1 μm to 10 μm, Functional Layer thickness is 2.5 μm)

electronic eye etc. The main advantage will be that these systems will sense and will take decision themselves like present drug delivery system for diabetic control.

Acknowledgements

Author is thankful to Dr. Pawan Kapur, Director, CSIO, for his continuous support. I am also thankful to Sh. N.K. Sharma at CSIO (Delhi Centre) and all my collaborators in this work, specially to Sh. C.R. Jalwania of SSPL. In preparation of the manuscript, help given by Mohammad Aziz Maalik is also acknowledged.

References

- Akar, O., Akin, T., Harpster, T., Najafi, K.; (2000). A wireless batch scaled absolute capacitive pressure sensor. EURUSENSORS XIV 585 – 588.
- Datskos, P.G., Thundat, T., Lavrik, N.V.; (2004). Micro & Nano Cantilever sensors, Encyclopedia on Nanoscience & Nanotechnology Ed. H.S. Nalwa Vol. X 1-10.
- Eggers, T., Marchner, C., Marschner, U., Clasbrummel, B., Laur, R., Binder, J.; (2000). Advanced Hybrid Integrated low power Integrated pressure monitoring system for biomedical applications, MEMS 2000 Miyazaki Japan.
- Ganesan, M., Sharma, N.K., Jain, V.K.; (2005). Optimisation of polymer coated silicon Microcantilever for chemical sensing applications, communicated for publication.

- Gattiker, G.E., Kaler, V.I.S. Karan, Mintchev.; (2005). Electronic Mosquito – A Semi –invasive micro – electrochemical system for blood sampling, Resume page 1–3.
- Goosen, JFL., (2002). Design Considerations for Silicon Sensors for use in Catheters and guide wires, Smart. Mater. Struct. **11**: 804–812.
- Jain, V.K., Jalwania, C.R., (1999). Development of IR Sensor using MEMS Tech. Int. Conf on Smart Mater., Structure and Systems ISSS, July 190–198.
- Jain, V.K., Jalwania, C.R., Swarankar, N.K., Singh, R.; (2000). Micro-flow meter integrated with an optical sensor SPIE **3321:** 758–770.
- Jain, V.K., and Jalwania, C.R.; (1998). Recent Trends in Smart Technology. Defence Science Journal **48 (1):** 21–29.
- Jain, V.K., Jalwania, C.R., Narang, R., Goel, T.C., Chatterjee, R.; (2004). Catheter End Temp. Probe for determining the temp. profile of arterial plague. Int. workshop on Smart Technology & Applications 81–82.
- Ji, Jin., Cho, S.T., Zhang, Y., Najafi, K., Wise, K.D.; (1992). An Ultraminiature CMOS Pressure Sensor for a Multiplexed Cardiovascular Catheter, IEEE Trans. ED **39(10):** 2260–2267.
- Kazunari, O., Aoyagi, S., Isono, Y., Hashiguchi, G., Fujita, H.; (2001). Fabrication of a Micro needle for a trace blood test, Transducers 01 Eurosensors XI, 412–415.
- Khaled, A.R.A., Vafai, K., Yang, M.; (2003). Sensors & Actuates B **94**: 103–115.
- Lin, L., Pisano, A.P.; (1999). Silicon – Processed Microneedles, IEEE J. Microelectromechanical System **8(1):** 78–84.
- Nagapriya, K.S., Raychaudhuri, A.K., Jain, V.K., Jalwania, C.R., Kumar Vikram; 2003, Effect of ambient on the Thermal Parameter of A Micromachined Bolometer **82(16)**: 2721–2723.
- Pei, J., Tian, F., Thundat, T.; (2004). Glucose Biosensor based on the Microcantilever, Anal. Chem. **76**: 292–297.
- Rasmussen, P.A., Thysen, J., Hansen, O.; (2003). Ultramicroscopy **97**: 371 – 376.
- Sorrell, J., Bhattacharya, A.; (1998). Mounting of MEMS Sensors on Coronary Guide Wires, University of Colorado at Boulder.
- Stefanadis, C., Toutouzas, K., Tsiamis, E., Stratos, C., Vavurankis, M., Kallikazaros, I., Panagiotakos, D., Toutouzas, P.; (2001). Increase Local Temperature in Human Coronary Atheroscterotic Plaques: Journal of the American College of Cardiology, Vol. 37 No. **5**:1277–1283.
- Tohyama, O., Kohashi, M., Fukui, M., Itoh, H.; (1997). A Fiber-Optic Pressure Microsensor for Biomedical Applications, "Transducer 97" 1489–1492.

2

Indian Initiatives in MEMS and Related Technologies: A Survey

S. Nirmala
International Advanced Research Centre for Powder Metallurgy and New Materials (ARC-I), RCI Road, Balapur PO., Hyderabad-500 005, A.P., India
Tel: 0091-40-24441075, Fax: 0091-40-24442699
E-mail: nirmala@arci.res.in

ABSTRACT

Micromachined Electro-mechanical systems (MEMS) also known as microfabricated systems (MS) have evoked great interest in the scientific and engineering communities all over the world. This is mainly due to a number of advantages that MEMS offer such as substantially reduced orders of magnitude in size, improved performance, possibilities for batch production and cost effective integration with electronics, almost zero power consumption etc. Realizing these advantages, many Indian scientific organizations have launched new activities on nanotechnology, smart materials, MEMS and its related fields. The Government of India funding bodies like Department of Science and Technology (DST) along with other departments have taken keen interest in promoting these technologies for new applications. The current field of Indian nanoscience and technology covers a wide spectrum of topics including MEMS, nanostructured materials synthesis and consolidation, DNA chips, drug delivery systems, nanoelectronics, pressure sensors, bio-sensors, gyroscopes, accelerometers and many other applications.

Most of the institutes concentrate on novel design aspects using these advanced technologies which includes Indian Institute of Science, Central Electronics Engineering Research Institute, Central Scientific Instruments Organization, Indian Institutes of Technology, Solid State Physics Laboratory, Tata Institute of Fundamental Research, Jawaharlal Nehru Centre for Advanced Scientific Research and many universities around India. In addition, manufacturing facilities for commercial exploitation exists at Central Electronics Engineering Research Institute, Indian Technical Institute, Bharat Electronics Limited, and Semiconductor Complex Limited etc. As an outcome of the spurt activities in India, even private institutions have joined the foray with many target-oriented applications for MEMS and related areas. The paper will present a review of state-of-the-art capabilities in MEMS and other related technologies in India.

Keywords: Nanotechnology, MEMS, Smart Materials, Key institutions, Fabrication, Sensors, Applications

Introduction

Micro-Electro-Mechanical Systems (MEMS) refers to the integration of mechanical elements, sensors, actuators and electronics on a common substrate through the utilization of micro fabrication technology. MEMS promises to revolutionize almost every product, as it is an enabling technology allowing the development of smart products for smart applications. Usually, the activities in MEMS are governed in the following ways: the sensors gather information from the environment through the measurement of mechanical, thermal, biological, chemical, optical and magnetic phenomena; the electronics process the information derived from the sensors; and through some decision making capability, the actuators are directed to perform various events by moving, positioning, regulating, pumping and filtering, enable controlling the environment for the desired purpose. Most of the MEMS devices are built on silicon wafers, where the intended components are formed in the form of thin films and using photolithographic methods, the markings are made to eventually print the design. These methods are similar and an integral part of conventional IC manufacturing.

Since MEMS devices are manufactured using batch fabrication techniques, similar to ICs, unprecedented levels of functionality, reliability, and sophistication can be placed on a small silicon chip at a relatively low cost. MEMS technology is enabling new discoveries in science and engineering such as the Polymerase Chain Reaction (PCR) microsystems for DNA amplification and identification, the micromachined Scanning Tunneling Microscopes (STMs), biochips for detection of hazardous chemical and biological agents, and microsystems for high-throughput drug screening and selection. In the industrial sector, MEMS devices are emerging as product performance differentiators in numerous markets with a projected market growth of over 50% per year.

A useful operative definition describing the scope of MST (Micro System Technology) is "Microstructural products having structures in the micron range and have their technical function governed by shape of the microstructure". In general, a microsystem combine several micro-components, which are optimized to behave as a single system providing one or several specific functionalities, including microelectronics applications. As evidenced by this definition, microsystems do not necessarily include integrated circuits, nor, are always monolithically integrated. The futuristic applicability of such microsystems are deemed to have a pervasive influence in many industrial sectors ranging from domestic to aeronautical, automotive and medical sectors. Some of the 'smart' products derived out of MEMS technology having noteworthy impact on future technologies is listed in Table 2.1.

Table 2.1: List of Smart Products Using MEMS

Examples of Smart Product	*Smart Uses*
Shape-memory alloy	A smart material
Building with an active damping system	A smart structure, such as a building that responds to changes in its surroundings
Pressure sensors	A smart sensor, like those used in automobiles
Micromotor	A smart actuator
Fuzzy controller	A smart controller
Neuronal chip	A smart electronics with intelligence
MEMS or MOEMS	A smart microsystem that comprises sensor processor and actuator, all integrated in one chip

MEMS Research Initiatives in India

Nanotechnology Initiatives

India is making its way to Nanotechnology era in a big way, as it is one of the two technologies identified by scientist turned President of India, Dr. A.P.J. Abdul Kalam himself. Nanotechnology finds applications in different areas such as new materials, manufacturing, chemical, space and energy industries besides medicine and healthcare. Most of the research activities on nanoscience is being conducted at some of the premier educational institutions like Indian Institute of Science, Bangalore; Indian Institutes of Technology in Chennai, Kharagpur, Mumbai, Delhi and Roorkee, Universities of Delhi, Hyderabad, Pune, Madras, etc. In addition, many government funded R & D laboratories International Advanced Research Centre, Hyderabad, Raman Research Institute, Bangalore, National Chemical Laboratory, Pune, Central Glass and Ceramic Research Institute, Jadavpur, have also initiated coordinated nanoscience and technology research focused on diverse areas.

In 2003, Department of Science and Technology initiated the nanoscience research program earmarking US$400,000 for research equipment alone, which was raised to US$26 million the following year. In continuation with the constant support and realizing the importance of nanotechnology, the funding was raised to US$50 million for the year, 2005 towards nanotechnology research almost doubling the initial allotment. However, in comparison with the funding of developed nations like US ($300 billion), China ($300 million), India's investment is rather modest.

Overview of MEMS Development in India

Primarily, due to the higher cost involvement in the basic infrastructures on MEMS, research activities were initiated by only government-funded institutes in India. All the diverse activities were covered under National Program on Smart Materials (NPSM) and Development Initiative for Smart Aircraft Structures (DISMAS). The NPSM aims to generate Indian capabilities in MEMS and smart systems and is being jointly sponsored by various government bodies like Council of Scientific and Industrial Research (CSIR), Defence Research and Development Organization (DRDO), Department of Science and Technology (DST), Indian Space Research Organization (ISRO) and Department of Information Technology (DIT). A key motivating factor for MEMS research in India has been to develop indigenous capabilities to meet the country's future requirements in strategic sectors such as defence and aerospace. However, there is also sizeable market in India for MEMS-based devices in the non-strategic industries like automotive, chemical, medical and pharmaceutical industries. In India, private companies have largely shied away from investing in research. However the CranesSci-MEMS lab is India's first privately funded MEMS Research Laboratory, which is a joint venture between Indian Institute of Science (IISc), Bangalore and Cranes Software International Limited (CSIL) also based in Bangalore. This augurs well for the future. It is hoped that more private companies will join in this national effort in the near future.

MEMS system includes mechanical and electrical elements that convert one form of energy into another, operating by transudation. MEMS devices are made of extremely small parts known as microchips. This miniature device is made up of mechanical elements, actuators and electronics on a common silicon substrate, fabricated using microsystems technology. MEMS is an assembly of parts having an association, such as a micro sensor-a miniature device that converts a non-electrical quantity like pressure, temperature, gas concentration or magnetic phenomenon - into an electrical signal and an actuator which converts an electrical signal into mechanized action. Another type of device is the Micro-Opto-Electro-Mechanical Systems (MOEMS), which offers higher bandwidth, lower cost, smaller size and provides relatively easier integration compared to similar optical devices in the macro world. All components of the MEMS devices are essentially to be integrated on a single platform, *i.e.*, the components are fabricated using same substrate. Since the issues like compatibility with the electronics controls are vital, mostly these components are fabricated using micro-machining and stereo-lithography, etc. Usually, they are manufactured in bulk as a batch production. There are number of technical issues involved during the design and development such as operating principle, microsystem process modeling, circuitry design, process technology, environment during application, packaging and manufacturing system. These issues are well realized in all the Indian research institutes to bring out high quality products.

Overview of MEMS Applications

Some of the application areas provided by MEMS devices include, silicon-based pressure sensors for measuring exhaust emissions, fuel, tyre and hydraulics in the automotive industry; polymerase chain reaction (PCR), microsystems for DNA amplification and identification (NEMS), biochips for identifying hazardous gases in the environment (NEMS), the electronic nose, with an inbuilt array of electronic chemical sensors as well as a pattern-recognition system, capable of recognizing simple to complex odours or an electronic tongue for sensing different tastes. MEMS devices used for automobile safety include accelerometers used for detection of collision and inflating an airbag, tyre-pressure monitoring and non-skid brake adjustments. Miniature satellites, of dimensions less than 10 cm across, microprobes and micromachined adaptive optics that allow correction of lens distortions, navigation and guidance systems in spacecraft propulsion are some of the applications of MOEMS.

The medical arena has microprobes with recording tips to measure biopotentials, smart inhalers that monitor droplets inhaled by a patient, catheters with smart sensors, sensors for locally measuring temperature and pressure, micro-needles and sensors for insulin drug delivery. Other examples concerning healthcare are pacemakers for adjusting pulse rate and implantable micro-pumps for delivering micro doses of medicine. It has to be mentioned that in many applications related to biological and medical fields, an advanced concept of nanosized NEMS systems are preferred over the conventional MEMS.

Fabrication Facilities for MEMS

Fabrication of MEMS structure includes the following steps:

1. Preparation of high purity silicon wafers with precise levels of doping.
2. Oxidation patterning of the wafers by lithographic techniques, etching, thermal annealing and thin film deposition.
3. Micromachining for removing part of the substrate or thin film either isotropically or anisotropically.
4. Bonding and packaging

Most of the above steps are common to manufacturing of VLSI and VLSI based systems.

Semiconductor Complex Limited, Chandigarh

It was established in 1983, with focus areas being design, development and manufacturing of VLSIs and VLSI based systems. Based on the previous records, SCL was identified as a national facility for MEMS manufacture in 2003. As a national foundry for MEMS, SCL provides process development support and manufacturing facility for product designs conceptualized by various design houses in the country. SCL has established primary processes and equipments for fabricating a wide variety of products such as accelerometers, pressure sensors for industrial, health and strategic applications, IR sensors etc.

Basic facilities available at SCL include,

1. Bulk Micro-machining,
2. Surface Micro-machining,

3. Wafer Bonding, and
4. MEMS Packaging

Using the above facilities, SCL activities include,

- Surface Micro-machining process with structural layers of silicon dioxide, poly-silicon and aluminium.
- Bulk micro-machining process with electrochemical etch stop and timed etch stop.
- Sub-micron processing (up to 0.8 microns minimum feature size) with I-line (365 nm) lithography.
- Front-to-back alignment.
- LPCVDE and PECVD processes for silicon dioxide (undoped, PSG and BPSG), silicon nitride and poly-silicon deposition.
- Metal deposition by sputtering for aluminium and titanium.
- Metal deposition by e-beam evaporation for gold, chrome, etc.
- RIE for metals, silicon nitride and silicon dioxide.
- Wet etch stations for metal, poly-silicon, silicon nitride, silicon dioxide and anisotropic etching of silicon.
- Medium and high current implanters for B+, BF2, P, As implantations at different values of dose and energy.
- Anodic and fusion bonding for Si-glass and Si-Si bonding.
- Characterization facility with non-contact surface profiler, SEM, semiconductor process control system, semiconductor parametric analyzer and various reverse engineering processes.
- Parametric testing with digital and analogue tester.
- Dicing process for the fragile MEMS structure.
- Packaging in ceramic (DIP and PGA) and COB in various lead counts.

Bharat Electronics Limited (BEL), Bangalore

Another fabrication center for MEMS has been established at Bharat Electronics Limited (BEL) at Bangalore. BEL is one of the leading fabricators of semiconductor components in India and is well equipped with the necessary infrastructure for manufacturing MEMS and semiconductors. BEL has three basic facilities:

- Process Facilities
- Design Facilities
- Assembly and Characterization Facilities

Process Facilities

Arrangements for the generation of ultra-pure gases such as oxygen, hydrogen, nitrogen as well as demineralized water connected on-line to the silicon wafer-fabrication and assembly units; the wafer fabrication units comprising of acid-processor cleaning, pyroginic oxidation and medium current ion implantation for boron, phosphorus and antimony, housed in class 100 and class 1000 clean rooms as per requirement; photo process equipment such as automatic coating facilities, mask aligning, exposing facilities, electron-beam metallisation as well as DC and RF sputtering facilities for various metal deposition. Additionally MEMS process equipment such as Double-side MASK, Bond Aligner, Anodic (Fusion Bonder, LPCVD Nitride, Polysilicon, RIE facilities have been added recently.

Design Facilities

The semiconductor wafer fabrication is supported by state-of-the-art software like CADENCE, SYNPOSYS, ADS and Microwave office. Facilities for conversion of CDS II and other layout formats to make chrome as well as emulsion hard masks and for fabrication of working masks with feature sizes down to 3mm are available. Recently Conventor Ware Software tool for simulating MEMS structures has been added.

Assembly and Characterization Facilities

These include device characterization facilities such as standard IV, CV and wafer level probe station; calibration facilities for low-pressure and high-pressure sensors upto 600 bar; accelerometer characterization test bench.

Assembly facilities include wafer dicing, die bonding, wire bonding and welding, molding and plating facilities.

In addition to the above institutes, well-equipped fabrication facilities are available at Central Electronics Research Institute, Pilani and at Indian Technical Institute, Bangalore.

Indian Institute of Technology (IIT), Kharagpur

At IIT Kharagpur, an Advanced Technology Centre (ATC) was established in July 1998. Conceived to grow in a modular fashion, the centre aimed to consolidate and strengthen research in the interdisciplinary areas of microelectronics, MEMS & microsystems, nanoelectronics and photonics. The Centre was equipped with advanced processing and characterization facilities like mask making & photolithography, silicon wafer processing, ion-beam deposition and etching, magnetron sputtering, MOVPE, PECVD and molecular beam epitaxy (MBE) systems, ellipsometer, surface profilometer, integrated optic testing setup, optical spectrum analyser, etc.

In the year 2000, creation of additional facilities and a MEMS Design Laboratory to expand MEMS and Microsystems activities was initiated. Clean rooms and silicon processing and testing facilities in the Microelectronics Laboratory were created from the special grant of the Ministry of HRD under the thrust area of emerging technology. Competence has been built up for the design and fabrication of silicon bipolar devices and integrated circuits. The chips developed in the laboratory include linear bipolar array, four-quadrant analog multiplier, ECL gate and relay driver chip.

The technology of ECL - the fastest bipolar logic, developed for the first time in the country, has received special attention. A sponsored project has been under taken to develop fast bipolar ICs for hazardous environment. A close techno-commercial link with Bharat Electronics Ltd., has been established for batch fabrication of some of the chips. Semiconductor Complex Ltd., Chandigarh is now helping in commercialisation of MEMS products.

Activities have also been launched on micromachining of silicon and quartz and developing special MEMS components. Several government departments including IISU/ISRO, ADAINPSM have funded projects to develop microsensors for special applications. Recently, the MEMS Design Laboratory, a national facility under NPSM programme has been setup to boost MEMS research activities in the country. Techniques have been developed for fabricating silicon and quartz microstructures like membranes, cantilevers, nozzles, miniature vibrators and tuning forks. Miniature piezoelectric quartz vibrating beam accelerometer has been successfully developed under an ISRO sponsored project. MEMS piezoresistive accelerometer is under development for avionics applications.

As mentioned earlier, there are also many design centers (especially all the academic institutes have their own design centers) spread all over the country catering to different applications.

Major R & D achievements in last five years in MEMS are:

- First ECL Gate and Bipolar Array Chips developed in the country
- Developed first Micro-machined Quartz Gyrochip
- First Quartz Accelerometer DETF Chip
- Silicon accelerometer for aircraft motion sensing

Major in-house facilities for MEMS are:

All related facilities like Clean Rooms, Silicon wafer processing, Pattern Generator, Photolithography, Diffusion/Oxidation, UHV electron-beam, Ion-beam system, Wafer level testing, Laser Writer, Double side mask aligner MA6 etc., silicon and Quartz micromachining, Integrated Optics Fabrication.

Design Centres

In India, there are many 'design centers' set up at different institutions catering to MEMS design. These are located at Indian Institute of Science, Bangalore, Solid State Physics Laboratory, Delhi, and Indian Institutes of Technology at Chennai, Mumbai, Kharagpur and Roorkee. The special feature of these design centers is the use of the 'same platform, same software for exchange of design' so as to avoid unnecessary duplication of efforts.

Central Glass and Ceramic Research Institute (CGCRI), Kolkata

PZT actuators based on piezoelectric effect are being made along with multi-stacking of actuators.

National Chemical Laboratory (NCL), Pune

Manufactured has made magnetostrictive materials for use in aerospace applications for vibration control, improving pointing accuracy of air borne antennae and for sensing damages in laminated compositions. These materials give 80% performance of expensive imported material with brand name developed magnetic sensor materials which given significant current under applied magnetic field.

Participation of Private Organizations

A MEMS research laboratory–Cranes Sci MEMS–has been set up with joint collaboration between the Bangalore based Indian Institute of Science and an Indian Software Company, Cranes Software International Limited. This research laboratory has been started with an initial grant from India's University Grants Commission, a government agency that helps to develop educational institutions. The Cranes Sci MEMS Laboratory will conduct design and testing, utilizing different fabrication facilities in India. The laboratory started in November 2001, has already filed its first patent on a MEMS microphone design. Having inherent technical strength in acoustics, vibrations, dynamics and digital signal processing the laboratory is strategically focusing on developing acoustic sensors, inertial sensor, ultrasound sensors as well as software design. The laboratory's long-term prospects include strategic sensor technology and medical diagnostic technology.

In addition, a few privately owned companies in India have either started work or are gearing up to take up MEMS seriously. These include, US based companies having operations in India such as Honeywell and General Electric as well as Indian companies Cranes Software and Titan Industries Limited. Due to many proven applications, Indian automotive industries are getting the products designed and developed by the existing research institutes. In due course of time, these automotive manufacturers are also expected to have their own spin-off company.

Other Related Initiatives in India

Smart Technology that includes MEMS and MOEMS has attracted International attention in recent years. After the International Conference held in Bangalore in December 1996, on smart materials, structures and systems, the Institute of Smart Structures and Systems (ISSS) was formed as a society comprising of all those working actively or interested in promoting smart technology in India. The ISSS was responsible for the formation of the National Programme on Smart Materials (NPSM). Under the NPSM devices and materials as well as the user agencies were identified. The Government of India sanctioned in July 2000 a fund of Rs.75 crores (US$15 million) spreaded over five years. NPSM is a joint programme run by Defence Research and Development Organization (DRDO) through the Aeronautical Development Agency (ADA), Bangalore along with Department of Space (DOS), Department of Science and Technology (DST), Ministry of Information Technology (MIT) and the Council for Scientific and Industrial Research (CSIR). A number of projects have been sanctioned by NPSM to promote the setting-up of facilities for design, fabrication, integration and qualification testing for MEMS related projects in order to strengthen and accelerate efforts, get the participation of industries, keep pace with rapid progress in this emerging field. The list of projects funded by NPSM is given in Appendix I.

Another programme is named Development Initiative for Smart Aircraft Structures (DISMAS) specifically for aerospace applications. DISMAS is also a five year project sanctioned by DRDO at a cost of Rs.19.26 crores (US$4 million) in April 2001. The project is being overseen by ADA, Bangalore and their areas are the following:

- Structural health monitoring involving monitoring of internal damage of laminated composite structures used in wings, fins and fuselage shell of aircrafts by smart sensors.
- Vibration and noise control in structural components of aircraft using flutter control, distributed sensors and actuators and damping enhancement systems.
- Active flow or shape control by changing the exterior contour of an aerodynamic surface for better aerodynamic effects with sensors and distributed actuators.
- Conformal antennae and radome actuators for making the external contour of the aircraft aerodynamically sound.

Smart technology can be used in damage monitoring of airframes made of advanced carbon-fibre composites, better manoeurvability of aircraft, suppression of vibrations in the fin tip and control surfaces. These concepts are to be tested using the Light Combat Aircraft (LCA) as a platform and could form a basis for incorporation of smart-technology in future models of LCA.

Smart technology products such as shape memory alloys can be used in aerospace applications - embedded in composites, nickel-titanium shape memory alloy can dampen vibrations - in automotive applications as fuel control valves, in medical field as surgical equipment, as actuators for manipulators in robots and in micro pumps, in the form of wires and ribbons for triggering, drive and as final control elements as well as valves and injectors. National Aerospace Laboratory (NAL), Bangalore has been successful in memory alloy fabrication, wire drawing with controlled microstructure.

Relaxor ferroelectric materials like lead magnesium niobate - lead titanate are useful as actuators. At C-MET, Hyderabad actuators based on multilayer relaxor materials are being made. Other projects in advanced stages are ferroelectric thin films on MEMS structures for developing RF MEMS used in microwave applications at the University of Hyderabad; nano porous materials like nano-titania integrated onto MEMS for use as chemical sensors at Indian Institute of Technology, Kharagpur; Multilayer - stacked ferroelectric - relaxor based actuators at National Aerospace Laboratory, Bangalore.

Smart technology with its extensive applications has been identified as an important tool for development in India.

Indian MEMS Devices–Selected Examples

(i) MEMS based SCL sensors (SP01, SP02) from Semiconductor Complex Limited (SCL): These are relatively small, low cost and reliable. These pressure sensors contain sensing element consisting of four-strain sensitive piezo-resistors ion implanted into silicon. These resistors are connected in a Wheatstone bridge configuration where resistance of two resistors increases, while other two decreases when pressure is applied to the sensor. The resistance in the arms of the bridge changes by an amount proportionate to applied pressure resulting in change of sensor output voltage. After the resistors are formed on the sensing chip. A diaphragm is created by chemically etching the silicon from the backside. The diaphragm thickness determines the pressure range of the sensor.

Features of Pressure Sensor, SCP01

- Absolute, Gauge or Differential
- Pressure Range: 0 to 1 Bar
- Excitation: 1V/3V DC or 1mA DC Current
- Full Scale Output: 200mV
- Package Type: TO-8 Metal Can Package
- Linearity: Better than 1%

Applications

BP Monitors, Meteorology, Pressure Transmitters, Digital Pressure Gauges

Pressure Sensor SCP01

Pressure Sensor SCP02

Features of Pressure Sensor, SCP02

- Absolute or Gauge
- Pressure Range: 0 to 10 Bar
- Excitation: 3V DC or 1mA DC Current
- Full Scale Output: 200mV
- Package Type: Custom Package
- Linearity: Better than 1%

Applications:

Tyre Pressure Measurement, Automatic Tyre Inflater, Digital Pressure Gauges, Pressure Transmitter

ii) MEMS devices, which are currently under development at Bharat Electronics Limited (BEL), Bangalore include flow sensing devices (pressure, gas), inertial sensing devices (accelerometer and gyroscopes) and RF devices (switches, tunable filters, relays). Low-pressure sensor deployed in light combat aircraft, high G accelerometers, LPG gas sensors and RF- MEMS components are the successful products produced by BEL. Work is in progress to integrate flow sensing or inertial sensing along with RF-MEMS to realize smart systems

Conclusion

MEMS technology is the most promising technology today. It is also the first and the most vital step towards nanotechnology and smart technology. It calls for multidisciplinary teams whether it is product development or process development. Almost all the developed countries are putting more and more manpower, material and money to be the leaders. India, which had lost the agriculture, industrial, electronic revolutions in the past, clearly understood the importance of this revolution and has initiated in all fronts with the help of a sizeable scientific community.

Acknowledgement

The author is extremely grateful to Dr.G.Sundararajan, Director, ARCI for considering me worthy to participate in this International workshop. I am also indebted to Dr.Shrikant V Joshi for guiding me in preparing my presentation. I happily acknowledge the help provided by Prof. P.C. Angelo, Department of Metallurgical Engineering, PSG College of Technology, Coimbatore, India, Mr.Arun Joshi, and Mr.A. Shiva Kumar, ARCI, Hyderabad, India.

References

- A. V. Chavan and K. D. Wise, (1997). "A batch-processed vacuum-sealed capacitive pressure sensor", Proceedings of IEEE International Conference on Sold-State Sensors and Actuators, 1449-1452.
- C. T. C. Nguyen, L. P. B. Katehi, and G. M. Rebeiz, (1998). "Micromachined devices for wireless communications", Proceedings of IEEE. **86:** 1756-1768.
- Funded project titled "Design, Simulation, Development and Fabrication of Gas Sensor based on MEMS Technology".
- Funded project titled "Design, and Development RF-MEMS Components".
- Funded project titled "Development & Supply of Integrated Pressure Sensors".
- Funded project titled "Packaging of Pressure Sensors / Accelerometers based on MEMS Technology".
- K. E. Petersen, (1978). "Dynamic micromechanics on Silicon: Techniques and devices", IEEE Transactions on Electron Devices. **25:** 1241-1250.
- K. Natarajan, (2005). "MEMS Activities at Bharat Electronics", Micro and Nano 2005, Delhi, India.
- K. Natarajan, K. Srinivasa Reddy and Chari K. S, (2003). "Trends in Gas Sensor Technology and Applications", Proceedings of IWPSD, 1245-1249.

L. J. Hornbeck, (1991). Spatial Light Modulator and Method, U.S Patent 5061049.

N. J. Nielsen, (1985) "History of thinjet print head development", Hewlett – Packard Journal. **36**: 4-10.

N. Yazdi, F. Ayazi, and K. Najafi, (1998). "Micromachined inertial sensors", Proceedings of IEEE. **86**: 1640-1656.

P. F. Van Kessel, L. J. Hornbeck, R. E. Meier and M. R. Douglass,(1998). "A MEMS-based projection display", Proceedings of IEEE. **86**: 1687.

Ranjay Laha, B. Sriram, A. T. Kalghatki, K. Natarajan, and Sayanu Pamidighantam, (2005). "Design, Development & Characterization of Surface Micromachined Passive Components for Radar Applications", Proceedings of IRS 2005.

R. T. Howe, (1988). "Surface micromachining for microsensors and microactuators", Journal of Vacuum Science Technology B. **6**: 1809-1813.

APPENDIX I: List of Projects Funded by NPSM

Sl.No	Project Title	Project Institute	Project Status
(A)	**Automotive/Aerospace (Encompassing Civil & Mechanical)**		
1.	Active Noise Control Device	NAL Bangalore	Completed
2.	Pressure Sensor	IIT Chennai	Completed
3.	Accelerometer	IIT Khargpur	Completed
4.	Constitutive Modelling	IIT Chennai	Completed
5.	Magnetic Materials	NCL Pune	Completed
6.	NI-TI based Shape Memory Alloys	NAL Bangalore	Completed
7.	Piezoelectric Actuators	CG & CRI Calcutta	Completed
8.	MOEM Pressure Sensor	IISc Bangalore	Completed
9.	MEMS Microswitch	IIT Delhi	Completed
10.	Millimeter-wave antennas	IIT Delhi	Completed
11.	Magnetic Oxide	IISc Bangalore	Completed
12.	Integrated Piezoelectric	IISc Bangalore	Completed
13.	Electroactive Polymers	NCL Pune	Completed
14.	Development of PZT & PMN Materials and fabrication of Sensor/Actuator devices	NAL (MSD) Bangalore	Completed
15.	Structural Health Monitoring of prestressed concrete bridge girder	SERC Chennai	Completed
16.	Design & Development of MEMS capacitative pressure sensor chip	CEERI Pilani	Ongoing
17.	Development of plasma Etching processes for high aspect ratio structuring for MEMS application	CSIO Chandigarh	Ongoing
18.	Silicon MEMS Foundry services	SCL Chandigarh	Completed
19.	Development of magneto-electric composites for sensors	University College of Sciences, Deptt. of Physics, Osmania University	Ongoing
20.	Dev of multilayer PMN-PT electrostrictive actuator	C-MET & ARDE	Completed
21.	Design & Development of Low-Pass Filter based on MEMS technology	BEL Bangalore	Completed
22.	Packaging of Pressure Sensor and Accelerometer	BEL Bangalore	Completed
23.	FBGs & LBGs for str health monitoring	CSIO Chandigarh	Ongoing
24.	Active millimeter wave antenna	IIT Delhi, SSPL Delhi	Completed
25.	Nanoporous Titania films	IIT Kharagpur	Ongoing
26.	Ferroelectric thin films for RF MEMS	University of Hyderabad	Ongoing
27.	Tech Demonstration of Multilayer LTCC packages for MEMS appln	C-MET Pune	Ongoing

Contd...

APPENDIX I–Contd...

Sl.No	Project Title	Project Institute	Project Status
28.	Upgradation of MEMS Design Centres	IISc Bangalore, IIT Bombay, IIT Kharagpur, SSPL/IIT Delhi, IIT Madras, IITGauhati/Kanpur	Ongoing
29.	Running of Silicon MEMS Foundry	SCL Chandigarh	Ongoing
30.	NiTi Shape Memory Alloy thin films for micro-actuator application	IISc Bangalore	Ongoing
31.	Torsional MEMS Varactor	IISc Bangalore, SCL Chandigarh	Ongoing
32.	Vibratory MEMS Gyroscope	IISc Bangalore, SCL Chandigarh	Ongoing
33.	Active Array(8x8) antenna	IIT Delhi, CSIO Delhi	Ongoing
34.	Packaging of RF MEMS Components	BEL Bangalore	Ongoing
35.	Aircraft Structural Repair Technology using Smart Actuators	NAL Bangalore	Ongoing
(B)	**Bio-Medical**		
1.	Silicon Locket for cardiac diagnostics	IIT Mumbai	Projects conceived by NPSM was completed with private participation
2.	Development of catheter-end temperature probe for arterial plaque	IIT Delhi, AIIMS Delhi, SSPL Delhi	Ongoing
3	Development of bio-MEMS based micro clinical diagnostic kit for tuberculosis	CSIO Chandigarh	
4	Bio-MS for cardiac diagnostics	IIT Mumbai	Ongoing
5	Smart Vest	DEBEL Bangalore	Completed
6	Non-invasive blood glucose monitor	IIT Kharagpur	Ongoing
7	Silicon micropump for drug delivery & drug dosage	IIT Madras	Ongoing
8	Magnetically Actuated miniature polymer pump	IISc Bangalore	Ongoing

Development of MEMS Based Microfluidic Devices at the Research Centre for Electronics and Telecommunications: Indonesian Institute of Sciences

Goib Wiran

Microelectronics Laborato

Research Centre for Electronics and Telecommunicati

Indonesian Institute of Sciences, Indones

Phone: 62 22 250 4661; Fax: 62 22 250 46

Email: goib@ppet.lipi.go

ABSTRACT

In this paper, it will be described the development of MEMS based microfluidic devices at the Research Centre for Electronics and Telecommunication. The devices are a micromachined column for a miniature GC system and a microinjector for fluid handling system. The work on column miniaturisation at RCET has been started back in 1996 under a collaborative project (TIL – Targeted Institutional Links) with the University of South Australia. Several micromachined columns have been designed, fabricated, and characterized. Extensive theoretical calculations have been employed in the design phase, which signified the improved performance of a micromachined column over a conventional one. Based on this, a standard silicon micromachining technique has been employed in the fabrication of several miniature columns. In the case of the miniature injector system, the main aim is to allow a small amount of sample fluid to be handled and transported into another microfluidic device, such as the micromachined column described above. Although there are many commercially available injectors that can deliver sample volumes of the order of microlitres, the silicon-based microinjector reported here offers a number of advantages such as small size, precise sample volume, allowing mass fabrication to reduce cost per individual device, and possible integration with an existing microengineered GC modules.

Keywords: MEMS, Microfluidic, Micromachining, Miniaturization, Microcolumn, Microinjector

Introduction

Advances in microengineering technology have made possible the fabrication of microfluidic devices with unique and precise geometrical shapes. In particular, it is the advance in the etching technique which allows the creation of devices such as microcolumns on silicon with various cross sectional profiles: circular, rectangular, or triangular. For many years, such columns have been explored for various applications, such as in the study of fluid flow (Pfahler *et al.*, 1991), chemical separation (Jacobson *et al.*, 1994), and micro total analysis (μ-TAS). The main advantages include small size, low manufacturing cost (due to batch fabrication), minimum dead volume, and possible integration with microsensors and electronic circuitry.

Our particular interest is Gas Chromatography (GC). The working principle of this analytical technique is based on the interaction between the components of a gas mixture and a liquid or solid stationary phase inside the GC column. This interaction involves solution-dissolution (with a liquid stationary phase) or adsorption (with a solid stationary phase) process, which causes the individual components of the gas mixture to be separated (resolved) in time as the carrier gas (mobile phase) sweeps them through the column. Traditionally, GC uses packed columns having large internal diameters (ID). Column length can be up to several meters. However, with the introduction of capillary columns, it has been shown that column ID and column length can be greatly reduced while increasing its resolving power, speed of separation, and column efficiency (Trehy, Yost & Dorsey 1986). Better yet is a microengineered capillary column which has added advantages of being rugged and can be integrated with microengineered sensors and sample injectors, resulting in a high-speed miniature GC system.

Over the past few years, there have been a number of GC related works employing silicon micromachined columns. One of the earliest was the Stanford miniature GC (Terry, Jerman & Angell 1979), which employed a 1.5-m-long microcolumn coated with a liquid stationary phase. Using a separately batched fabricated thermal conductivity detector (TCD) and a microengineered sample injector, the column was able to separate a mixture of hydrocarbons in less than 10 s. However, peak resolution was somewhat poor due to non-uniformity of the liquid phase coating. A novel technique by vacuum sublimation of a CuPc thin film was then introduced to fix this problem (Kolesar, Jr. & Reston 1994). Although this technique has the advantage of being compatible with microengineering processes and results in a film with better adhesion properties, it still has some disadvantages. Firstly, this technique is restricted to gas-solid chromatography because of the nature of the deposited film, and secondly, it requires more processing steps such as cleaning up surfaces covered by unwanted film prior to the anodic bonding process. A newly developed method is by etching through the silicon wafer and oxidizing the resulting column walls to produce silica surfaces similar to those of fused silica capillary columns (Matzke *et al.*, 1998). After anodically bonded with a Pyrex glass, the column can be coated with a liquid phase using a conventional technique.

In addition to having a uniform stationary phase, column performance can also be improved by the use of a solid state microsensor and a microsample injector (Bruns, 1992). The fast time constant of the sensor and the ability of the injector to supply minute amounts of sample results in a high-speed micro GC system. In a situation where a very complex mixture is to be separated, column performance can be improved by using a multidimensional configuration (de Geus & Th. Brinkman 1996). Several columns can be coated with different stationary phases and then arranged in either a serial or parallel configuration. This method requires that the columns be highly versatile, and equipped with minimum (zero) dead volume interconnection fittings. Unfortunately, the design and fabrication complexity increases with increasing column performance.

Therefore, it is the objective of this work to develop a simple, flexible, and highly efficient microcolumn suitable for GC analysis. In addition to the microcolumn, a miniature injector is also developed to allow a small amount of sample fluid to be supplied into the microengineered GC column. The design, fabrication, and characterization of such micro devices will now be described.

Design and Fabrication

The block diagram of a typical miniature GC is shown in Figure 3.1. It usually consists of five basic components: carrier gas supply, sample injection unit, capillary column, gas detector(s), and electronic circuitry. The core of a GC system is the column, where separation of gas mixtures takes place. In a GC system using capillary column, a thin layer of adsorbing material, also known as the *stationary phase*, is usually coated over the inner walls of the column.

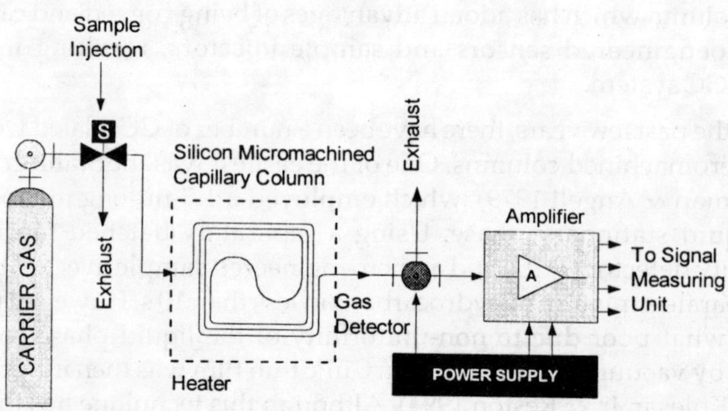

Figure 3.1: Block Diagram of a Miniature GC System

The operating principle of a GC system can be described as follows. First, an inert carrier gas such as nitrogen or helium is continuously delivered into the capillary column at a controlled rate. Then a small amount of gas sample is introduced into the carrier gas stream using the sample injection system. The sample is then injected and transferred by the carrier gas into the column for separation. Finally, the separated gas components are received and detected by the gas detector(s) at the end of the capillary column.

The Microcapillary Column

Design

In designing a microengineered GC, the first consideration is the length of the capillary column to be used. A common perception is that longer column is to be preferred as it gives a higher resolving power, but a shorter capillary column offers a number of advantages (Trehy *et al.*, 1986), and a compromise is therefore prompted (Wiranto *et al.*, 1999). The advantages of using a short capillary column include faster analysis time due to the increase in the optimum carrier gas velocity and smaller value of *theoretical plate height* due to a lower column pressure.

A layout of the micromachined columns is shown in Figure 3.2. As can be seen, the column has been designed as an interlocking spiral to minimize the total area occupied by the column tracks. Thus, with a column width of 100 μm and spacing between column tracks of 500 μm, a substrate area of 3.1 cm x 3.1 cm results in a column length of 125 cm. Compared with a square spiral, a shorter column would result if the column is designed in a circular pattern with the same dimensions. If column depth is expected to be around 20 μm after the etching process, the microcolumn should comprise a total volume of 2.5 μl. An important consideration in the design of the column is the expected pressure drop across the column. A large pressure drop requires high inlet pressure, which is undesirable because of the limitation given by the strength of the interconnection tubing attached to the column inlet/exit ports. Thus, the use of rounded corners was selected over sharp corners.

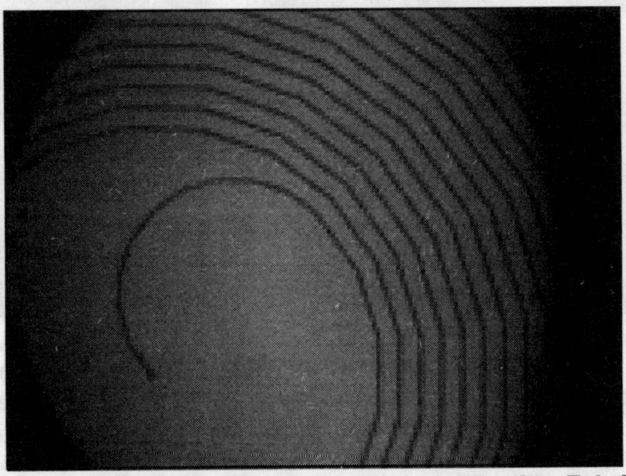

Figure 3.2: The Mask for the First Microchannel Design Fabricated Using E-beam Lithography

Fabrication

The starting material in the fabrication of the microcolumn was a 3-inch, (100), n-type silicon wafer. A 1.2-µm SiO_2 layer is thermally grown as a masking layer during the etching process. As has been described elsewhere (Wiranto *et al.*, 1997), isotropic etching technique was used to form the microcolumn grooves. The etching solution was composed of a 100 ml HNO_3 70% + 6 HF 48% + 40 H_2O. With this composition, the etching rate was expected to be slow as it was determined mainly by the small amount of HF (rate controlling). A slow etching rate would give a better control on the column depth. Following photolithographic patterning of the column tracks, the wafer was dipped in the solution for 10 min at room temperature, which resulted in grooves of approximately 14 µm deep, giving an estimated etch rate in the vertical direction of 1.4 µm/min. To obtain 20 micron deep grooves, the wafer was etched for a further 5 min. No agitation was used during the etching process. The resulting profile, as shown in Figure 3.3, is an almost-rectangular cross section with smooth wall surfaces and perfectly rounded corners.

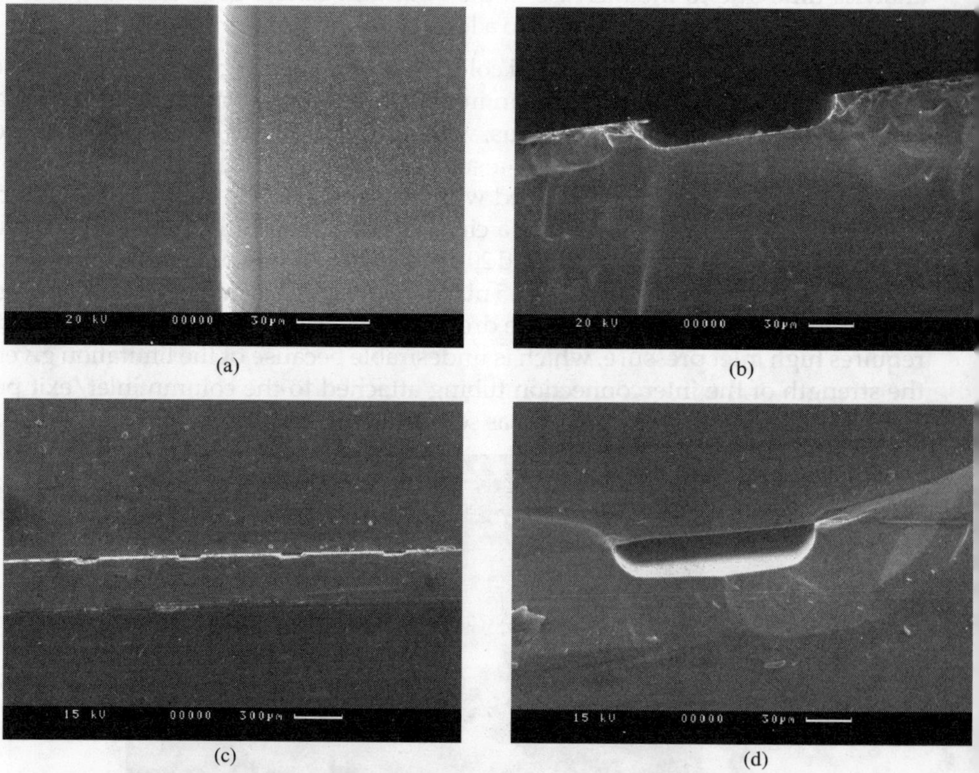

Figure 3.3: SEM View of Column Cross Sectional Profiles
(a) Top view before anodic bonding, (b) Front view before anodic bonding, (c) Far view after anodic bonding, and (d) Close-up view after anodic bonding.

To hermetically seal the column grooves, a Pyrex 7740 glass plate was anodically bonded with the etched silicon wafer. The choice of using a Pyrex glass is based on its thermal expansion coefficient, which is closely matched to that of silicon. In addition, Pyrex glass is chemically inert to most gases, and transparent, making it easier to visualize the bonding quality with the silicon. Prior to the bonding process, the glass is mechanically drilled to provide holes for the inlet and exit port connections. The bonding was performed at a temperature of 450°C using an applied DC voltage of 500-750 V. Although bonding takes place almost instantly at this temperature, the time required to completely seal the entire column area shown in Figure 3.1, is well over 4 hours. Some of the reasons for this long bonding time are due to the relatively thick Pyrex glass used (1.6 mm) and the low applied voltage.

A critical step in the column fabrication is tubing attachment at the column inlet/exit ports. The technique used of epoxy bonding has a high risk of the epoxy blocking the column. To reduce this risk, a low viscosity silver-based conductive epoxy (Epo-Tek H31D) was used. As shown in Figure 3.3, a short stainless steel tube (3-5 mm long) with an inner diameter of 180 µm was inserted into and attached to the ports. After curing at 150°C for 20 minutes, a fused silica tubing (150 µm OD and 50 µm ID) was put inside and attached to the stainless steel tubing to serve as a connecting tube in the coating process.

Column Coating

One of the most important decisions in the fabrication of the microcolumn is the choice of proper stationary phase. Requirements that should be considered when selecting the stationary phase material include its chemical stability, vapor pressure, viscosity at the operating temperature, and its selectivity and dissolving power for the components to be separated (Baiulescu & Ilie 1975). In addition, the stationary phase should have adequate wetting and adhesion with the column wall, in this case with silicon and Pyrex glass. Lack of adhesion results in a non-uniform coating, which will in turn degrade the efficiency of the column.

Although solid stationary phases have advantages over liquid phases, for example they can be deposited using microelectronic processing (Kolesar, Jr. & Reston 1994), the great majority of gas chromatography still uses liquid stationary phases. The main reason is that a wide range of liquid phases are available, thus making it easier to find selective phases for a particular separation.

To demonstrate the efficiency of the microcolumn, a non-polar liquid stationary (dimethylsiloxane) phase was applied to the microcolumn by means of a conventional dynamic coating technique (Concus 1970). With this technique, the stationary phase is dissolved in a volatile solvent, and a plug of this mixture is forced through the column. By maintaining a constant speed of the mixture through the microcolumn, a uniform layer of stationary phase is left coating the column walls. The resulting film thickness (d_f) can be calculated from the formula given by Novotny and Bartle (1974), which is expressed as:

$$d_f = \frac{rc}{200}\left(\frac{u\eta}{\gamma}\right)^{\frac{1}{2}} \quad (1)$$

where,

r is the radius of the column, *c* is the concentration of the liquid phase, *u* is the velocity of the coating plug, *hð* is the viscosity, and *gð* is the surface tension.

Characterization

The basic setup for the column characterization is illustrated in Figure 3.4. The column of a HP 5890 GC system was replaced by the microcolumn. One end of the fused-silica connecting tubes is fitted to the sample injector, whereas the other end is connected to the Flame Ionization Detector (FID). The output peaks of the FID are recorded and displayed on a central computer.

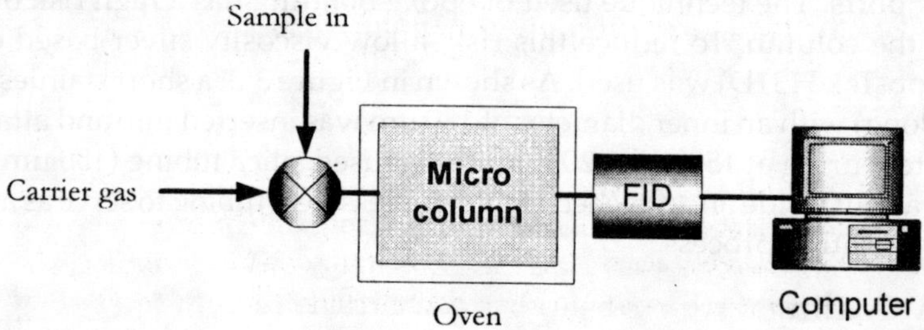

Figure 3.4: Experimental Setup for the Microcolumn Characterization

The test was conducted by injecting a 0.1 μl of sample containing hydrocarbon mixture (C_{11} to C_{16}) at isothermal (Oven) temperature of 150 °C. Helium was used as the carrier gas. Four different microcolumns were tested at various column inlet pressures, ranging from 29 to 70 psi. In order to make sure that the separation occurs only inside the microcolumn, the test for each column was performed at various conditions. For example, the initial test was carried out when both connecting tubes were still coated with the liquid phase. In other words, the same connecting tubes used in coating the column were re-used in this initial test. Next one of these connecting tubes was replaced and the second test performed. Finally, the remaining coated connecting tube was replaced and the third test conducted.

Results and Discussion

An advantage of using hydrocarbon samples in this work is that they have been known to be very soluble in non-polar stationary phases, and that they usually elute in the order of increasing boiling point (with increasing carbon number). This is shown in the chromatogram of Figure 3.5, obtained using one of the four tested microcolumns. As can be seen, at the given operating conditions (temperature and pressure) it takes about 1.6 minutes to completely separate the mixture using a column of 125 cm long. This chromatogram was obtained when both connecting tubes were free from liquid phase, which means that the separation occurred as a result of the coating inside the microcolumn only.

In general, the results obtained from the other columns showed that each column was able to perform separation of the hydrocarbon sample. However, the degree of

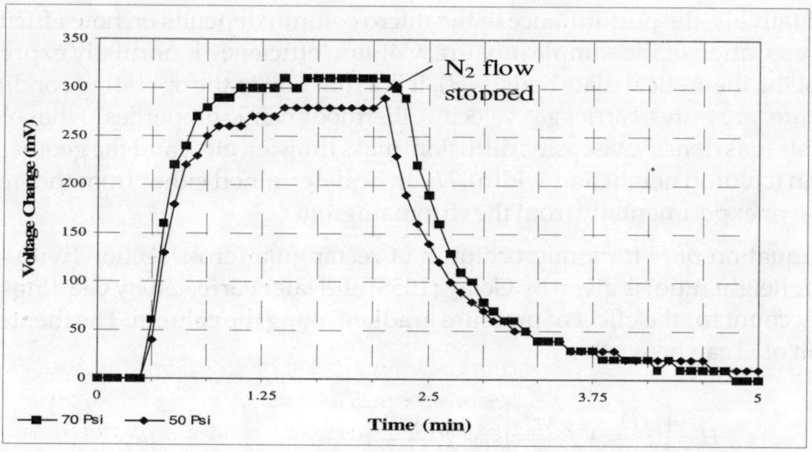

Figure 5. Recovery time of the detector under nitrogen gas in still air for two different inlet pressures: First steady state temperature = 175 °C

Figure 3.5: Chromatogram of Hydrocarbon Mixture (C_{11} to C_{16}) Separated Using a Microcolumn : Inlet Oressure 70 psi, Column Temperature 150°C, He Carrier Gas, Linear Velocity 30.22 cm/s

the separation varies for each column. As shown in Figure 3.6, these variations include peak tailing and poor resolution in some cases. We attribute these phenomena to the non-uniformity of the liquid phase coating. As noted in the previously published works (Kolesar, Jr. & Reston 1994), the uniformity of the stationary phase is very critical to the performance of a microcolumn. A more uniform film coating can be achieved by roughening the column walls, or otherwise a different coating technique should be employed. In this work, the column walls have been left unroughened.

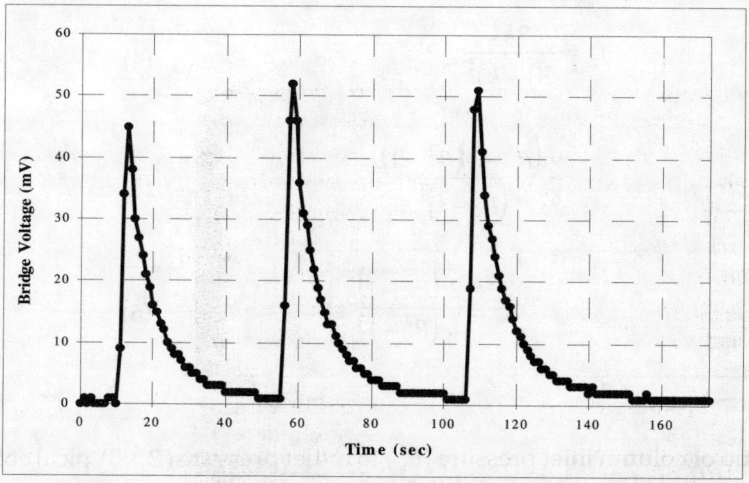

Figure 3.6: Chromatogram Showing an Unresolved Peak (C_{11}) Due to Non-uniformity of the Liquid Phase Inside the Microcolumn : Inlet Oressure 29 psi, Column Temperature 150ºC, He Carrier Gas, Linear Velocity 27.22 cm/s

Qualitatively, the performance of the micro column depends on how efficient it does the separation of the sample mixture. Column efficiency is normally expressed in terms of the theoretical plate height (H). It is a function of the operating conditions (temperature, pressures, carrier gas velocity), thermodynamic properties of the column components (gas density, viscosity, diffusion, mass transfer, etc.), and the geometry of the column (column height and width). H can be determined either from theoretical calculation or experimentally from the chromatogram.

The equation of H for a microcolumn of rectangular cross section (with large width over height ratio) is given by Golay (1958) and later corrected by Giddings *et al.* (1960) to account for the effect of pressure gradient along the column. The theoretical expression of H can be written as:

$$H = \left[2\frac{D_m}{v} + f_1 \frac{vz^2}{D_m} \right] g_1 + f_2 \frac{vz^2}{c^2 D_s} g_2 \tag{2}$$

where,

D_m is the diffusion coefficient in the gas (mobile) phase, D_s is the diffusion coefficient in the stationary (liquid) phase, v is the carrier gas velocity, f_1 and f_2 are functions determined by the partition ratio (k), c is the partition coefficient, z is half of the column height, and g_1 and g_2 are pressure gradient correction factors. The values of g_1 and g_2 are dependent on the ratio of pressures at the column inlet to outlet. The expressions for f_1, f_2, g_1, and g_2 are given respectively as

$$f_1 = \frac{4(1 + 9k + 25.5k^2)}{105(1+k)^2} \tag{3}$$

$$f_2 = \frac{2k^3}{3(1+k)^2} \tag{4}$$

$$g_1 = \frac{9}{8}\frac{(P^4 - 1)(P^2 - 1)}{(P^3 - 1)^2} \tag{5}$$

$$g_2 = \frac{3}{2}\frac{(P^2 - 1)}{(P^3 - 1)} \tag{6}$$

where,

P is the ratio of column inlet pressure (P_{in}) to outlet pressure (P_o). Typical column operational parameters are listed in Table 3.1. Given the column inlet pressure and linear velocity of the carrier gas, the expected (theoretical) value of H can be evaluated for each operating condition.

Table 3.1: Typical Values of Column Operational Parameters

Parameters	Symbol	Value	Unit
Column width	w	100	μm
Column half-height	z	10	μm
Column length	L	125	cm
Partition ratio	k	10	–
Diffusion coefficient in the mobile phase (Helium)	D_m	1.28	cm²/s
Diffusion coefficient in the stationary phase	D_s	10^{-6}	cm²/s
Thickness of stationary phase	d	0.2	μm
Column outlet pressure	P_o	1	atm
Partition coefficient	c	416	–

The chromatogram obtained in each test can be used to determine the experimental value of H for a specific sample component. Knowing the retention time (t_R) (the time measured from the start of injection to the maximum of the peak), the base width of the peak (w), and the column length (L), the expression is given as Desty *et al.*, 1957)

$$H = \frac{L}{16}\left(\frac{w}{t_R}\right)^2 \quad (7)$$

Assuming the peak is Gaussian, w is measured from the points of intersections of the tangent lines at the inflexion points with the base line.

Table 3.2 shows some comparisons of H values calculated using Equation (2) (Theoretical) and Equation (7) (Experimental) for the last eluted peak (C_{16}) obtained from various microcolumns tested. Test # 3 and # 4 are the same column but different column inlet pressures. As can be seen, increasing the inlet pressure did not alter the theoretical plate height significantly. This is in accordance to Equation (5) and (6), where the values of g_1 and g_2 will approach 1.125 and 0 respectively, as P_{in} gets larger. In most cases, the experimental values of H are slightly higher than the theoretical predictions. Again, this is primarily attributed to the non-uniformity of the liquid phase.

Table 3.2: Comparison Between Theoretical and Experimental Values of *H* for the Last Peak (C_{16}') from Various MOT Columns Tested

Test #	Theoretical H (mm)	Experimental H (mm)	Inlet Pressure (Psi)
1	0.63	0.57	50.7
2	0.80	0.87	70
3	0.96	1.45	29.1
4	0.96	1.43	38.7

The Miniature Injector System

The aim of the design of the miniature injector is to be able to supply a small amount of sample fluid into the microengineered GC column. The injected sample plug must be of nanolitre in volume and introduced in milliseconds. Too large a sample volume can cause column overloading, the effect of which on column performance is poor peak resolution. In addition, the injection system should allow the sample loop to be cleared by an inert carrier gas. Although there are many commercially available injectors that can deliver sample volumes of the order of microlitres (see, for example, the GC catalogue from Valco Instruments Co. Inc.), the silicon-based microinjector reported here offers a number of advantages. These advantages include: small size, precise sample volume, allowing mass fabrication to reduce cost per individual device, and possible integration with an existing microengineered GC modules.

System Design

The microinjector system has been designed to incorporate four one-valve injectors, as shown in Figure 3.7. The working principle of the microinjector systems can be described as follows. With valves #1 and #2 in off (closed) position, and valves #3 and #4 in on (open) position, a carrier gas entering from valve #3 is continuously flushing the GC column. When valve #1 is on, and valves #2, #3, and #4 are off, the sample gas injected through valve #1 will fill the short micromachined channel. The full capacity of this microchannel is approximately 4 nl. Once the microchannel is filled with the sample, valve #1 can be turned off and valve #3 opened, which results in the sample being pushed into the GC column. Alternatively, if valve #4 is now turned off the sample will be discarded to the system exhaust and the microchannel is cleared.

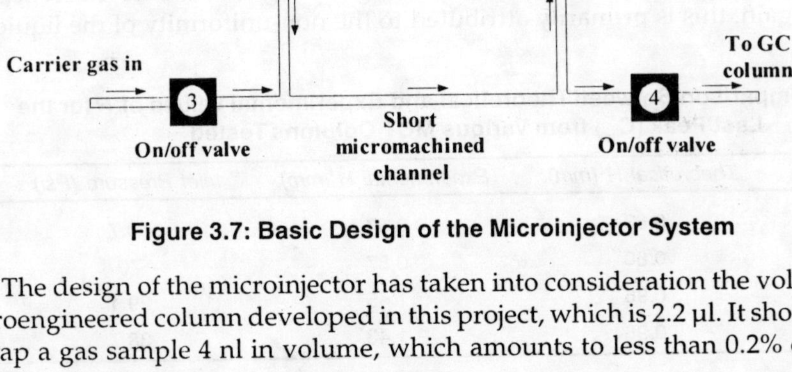

Figure 3.7: Basic Design of the Microinjector System

The design of the microinjector has taken into consideration the volume of the microengineered column developed in this project, which is 2.2 µl. It should be able to trap a gas sample 4 nl in volume, which amounts to less than 0.2% of the total column volume. As can be seen from Figure 3.1, the dead volumes associated with this design arise in the region between the microchannels after valves #1 and #2 and the main channel connecting valves #3 and #4. These dead volumes can be minimised by shortening the length of these channels as much as allowed by the dimensions of the valves. Another option is to reduce the width of the channels at these sections only.

The microinjector system has been designed to incorporate four one-valve injectors, as shown in Figure 3.1. The working principle of the microinjector systems can be described as follows. With valves #1 and #2 in off (closed) position, and valves #3 and #4 in on (open) position, a carrier gas entering from valve #3 is continuously flushing the GC column. When valve #1 is on, and valves #2, #3, and #4 are off, the sample gas injected through valve #1 will fill the short micromachined channel. The full capacity of this microchannel is approximately 4 nl. Once the microchannel is filled with the sample, valve #1 can be turned off and valve #3 opened, which results in the sample being pushed into the GC column. Alternatively, if valve #4 is now turned off the sample will be discarded to the system exhaust and the microchannel is cleared.

The design of the microinjector has taken into consideration the volume of the microengineered column developed in this project, which is 2.2 μl. It should be able to trap a gas sample 4 nl in volume, which amounts to less than 0.2% of the total column volume. As can be seen from Figure 1, the dead volumes associated with this design arise in the region between the microchannels after valves #1 and #2 and the main channel connecting valves #3 and #4. These dead volumes can be minimised by shortening the length of these channels as much as allowed by the dimensions of the valves. Another option is to reduce the width of the channels at these sections only.

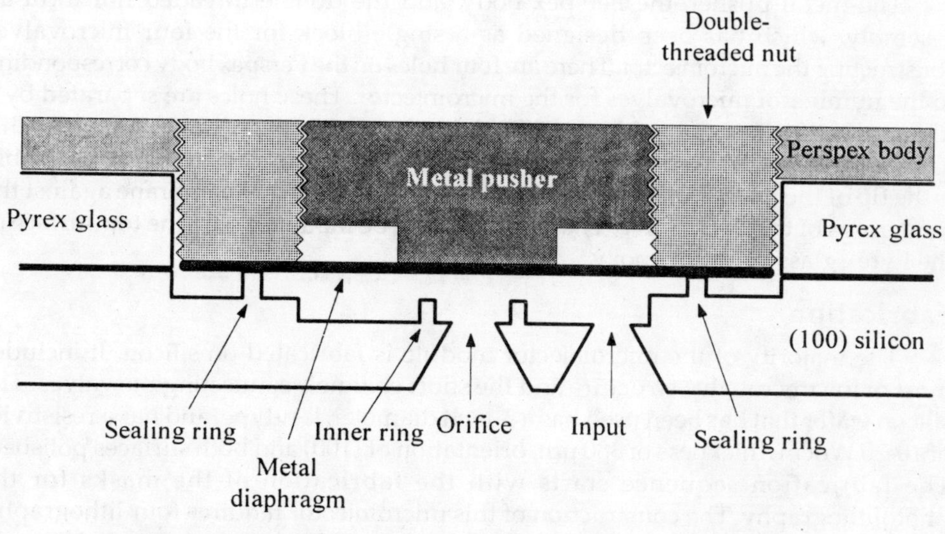

Figure 3.8: Basic Cross-sectional Structure of the Microvalves Used to Construct the Microinjector

Valve Structure

There are many choices for the on/off valve structures, based on those published in the literatures over the past few years (Esashi 1990; Shoji, Nakagawa & Esashi 1990; Shoji & Esashi 1994; Babaei & Kwok 1999). They can be divided into active microvalves (with an actuator) or passive microvalves (without an actuator). Each type has some advantages over the other. For example, an active microvalve allows its on and off positions to be controlled more easily using electronics, whereas a passive microvalve can be made extremely small and integrated with other on-chip components.

The on/off valve developed in this project is a simple, manually actuated diaphragm valve. The structure of this valve is a modification of that of Terry, Jerman and Angell (1979). As shown in Figure 3.8, it consists of an etched silicon ring and an orifice, a metal diaphragm, a (manually operated) metal pusher, and a Perspex body. Under normal condition, the metal diaphragm is released and the valve is open. When the threaded metal pusher is turned downward, the metal diaphragm deflects and is pushed against the inner rings of the valve. Under this condition, the valve will be closed and no gas can enter the orifice. Thus by turning up and down the threaded metal pusher, the valve can be operated in 'on' and 'off' positions.

The main component of the microvalve is the thin metal diaphragm. Besides strength, the main requirement for the diaphragm is inertness against various gases. In GC application, the injected sample mixture can be chemically reactive and corrosive. Hence, the chosen materials for the metal diaphragm must be able to withstand chemical attack and corrosion by sample gases. A nickel diaphragm seems to be a good choice for it is known to be non-corrosive and inert against a wide range of chemicals.

The metal pusher, the Perspex body, and the double threaded nut form an assembly, which has been designed as a single block for the four microvalves constructing the microinjector. There are four holes on the Perspex body corresponding to the number of microvalves for the microinjector. These holes are separated by a distance of 10 mm on one side and 15 mm on the other side. All components of this assembly have been fabricated using fine machining. A critical part of this assembly is the tip of the metal pusher, since it will be used to press the membrane against the inner rings of the microvalve. The assembly is to be mounted onto the top surface of the Pyrex glass using an epoxy.

Fabrication

The majority of the microinjector module is fabricated on silicon. It includes most of the microvalve structure and the short channels connecting the valves. The silicon wafer that has been used has a 3-inch diameter, is n-type, and has a resistivity of 10-20 Wðcm, thickness of 380 μm, orientation of (100) and both surfaces polished. The fabrication sequence starts with the fabrication of the masks for the photolithography. The construction of this microinjector requires four lithographic masks. Each mask has been designed using an AutoCAD 12 drawing tool. All masks were fabricated on a 4-inch chrome coated soda lime glass. Each mask contains four designs to maximise the utilisation of the 3-inch wafer area. The detail fabrication steps have been described elsewhere (Soegandi & Wiranto, 2004).

Conclusion

The authors have described the design, fabrication, and characterization of microcolumns suitable for GC analysis. Fabricated using microengineering technology, the column may be thought of as a flexible module which can be easily interfaced with an existing GC system for rapid characterization. In the future, several of these columns can be arranged in a multidimensional configuration using proper interconnection fittings. Coating of the liquid stationary phase remains the most critical step in the fabrication of the columns. A thin and uniform coating is necessary to guarantee a highly efficient column. In addition, improvement in column efficiency can be achieved by using a better connection technique at the column inlet/exit ports, one that minimizes dead volume. Based on the separation of hydrocarbon mixture, the performance of the column as a quantitative analytical device is more than satisfactory. The design of the microinjector for the miniature GC system has also been presented in this paper. It consists of four microvalves interconnected with short microchannels in such a way that a total of 4 nl sample can be injected into the microengineered GC column. The fabrication of the microinjector requires four lithographic masks, and is accomplished on a double sided polished (100) silicon wafer.

Acknowledgement

The works described in this paper has been partly sponsored by the Targeted Institutional Links (TIL) Project between Indonesian Institute of Sciences, Bandung Institute of Technology, and University of South Australia. Some fabrication steps were conducted at the Toyohashi University of Technology, Japan.

References

- Babaei, J. H. & Kwok, C. Y. (1999). 'Characteristics of a monolithic microvalve with large displacement electrostatic actuator', in *Electronics and Structures for MEMS*, N. W. Bergmann, O. Reinhold, N. C. Tien, Editors, Proceedings of SPIE, vol. 3891: 157-165.
- Baiulescu, G. E. & Ilie, V. A. (eds.) (1975). *Stationary Phases in Gas Chromatography*, Pergamon Press, Hungary.
- Bruns, M. W. (1992). 'Silicon micromachining and high speed gas chromatography', *Proceedings - IECON' 92, International Conference on Industrial Electronics, Control and Instrumentation*, vol. 3, San Diego, USA, 1640-1644.
- Concus, P. (1970). 'On the liquid film remaining in a draining circular cylindrical vessel', *Journal of Physical Chemistry*, vol. 74: 1818-1819.
- De Geus, H. J. and Th. Brinkman, U.A. (1996). 'Multidimensionality in gas chromatography', *Trends in Analytical Chemistry*, vol. 15: 168-178.

- Desty, D. H., Gleuckauf, E., James, A. T., Keulemans, A. I. M., Martin, A. J. P. & Phillips, C. S. G. (1957). 'Nomenclature recommendations', in *Vapour Phase Chromatography*, eds. D. H. Desty and C. L. A. Harbourn, Butterworths, London, xi-xiii.
- Esashi, M. (1990). 'Integrated micro flow control systems', *Sensors and Actuators*, vol. A21-A23: 161-167.
- Giddings, J. C., Seager, S. L., Stucki, L. R. & Stewart, G.H. (1960). 'Plate height in gas chromatography', *Analytical Chemistry*, vol. 32, No. 7: 867-870.
- Golay, M. J. E. (1958). 'Theory of chromatography in open and coated tubular columns with round and rectangular cross-sections', in *Gas Chromatography*, ed. D. H. Desty, Butterworths, London, 36-55.
- Jacobson, S. C., Hergenroder, R., Koutny, L. B. & Ramsey, J. M. (1994). 'High-speed separations on a microchip', *Analytical Chemistry*, vol. 66, 1114-1118.
- Kolesar, Jr., E. S. & Reston, R. R. (1994). 'Silicon-micromachined gas chromatography system used to separate and detect ammonia and nitrogen dioxide - Part I: Design, fabrication, and integration of the gas chromatography system', *Journal of Microelectromechanical Systems*, vol. 3, 134-146.
- Matzke, C. M., Kottenstette, R. J., Casalnuovo, S. A., Frye-Mason, G. C., Hudson, M. L., Sasaki, D. Y., Manginell, R. P. & Wong, C. C. (1998). 'Microfabricated silicon gas chromatographic micro-channels: fabrication and performance', in *Micromachining and Microfabrication Process Technology IV*, Proceedings of SPIE, vol. 3511, 262-268.
- Novotny, M. & Bartle, K. D. (1974). 'Preparation of thick-film glass capillary columns by the dynamic coating procedure', *Journal of Chromatography*, vol. 93, 405-411.
- Pfahler, J., Harley, J., Bau, H. H. & Zemel, J. (1991). 'Liquid and gas transport in small channels', *Micromechanical Sensors, Actuators and Systems*, ASME Winter Annual Meeting, New York, 49-60.
- Shoji, S. & Eshashi, M. (1994). 'Microflow devices and systems', *Journal of Micromechanics and Microengineering*, vol. 4, 157-171.
- Shoji, S., Nakagawa, S. & Esashi, M. (1990). 'Micropump and sample-injector for integrated chemical analyzing systems', *Sensors and Actuators*, vol. A21-A23 189-192.
- Soegandi, T. M. S. & Wiranto, G. (2004). 'The design of the miniature injector for the microengineered GC system', *Jurnal Elektronika dan Telekomunikasi*, vol. IV No. 1-2: 7 – 14.
- Terry, S. C., Jerman, J. H. & Angell, J. B. (1979). 'A gas chromatographic air analyzer fabricated on a silicon wafer', *IEEE Trans. Electron Devices*, vol. ED-26, 1880-1886.
- Trehy, M. L., Yost, R. A. & Dorsey, J. G. (1986). 'Short open tubular columns in gas chromatography/mass spectrometry', *Analytical Chemistry*, vol. 58, 14-19.

- Wiranto, G., Haskard, M. R., Mulcahy, D. E., Davey, D. E. & Dawes, E. E. (1999). 'Microengineered open tubular columns for GC analysis', in *Electronics and Structures for MEMS*, eds N. W. Bergmann, O. Reinhold, and N. C. Tien, Proceedings of SPIE, vol. 3891, 167-176.
- Wiranto, G., Samaan, N. D., Mulcahy, D. E. & Davey, D. E. (1997) 'Microfabrication of capillary columns on silicon', in *Smart Electronics and MEMS*, eds A. Hariz, V. K. Varadan & O. Reinhold, Proceedings of SPIE, vol. 3242, 59-64.

4

Development of Mems Based Microsensor Devices at the Research Centre for Electronics and Telecommunications

Totok M. S. Soegandi
Director
Microelectronics Laboratory,
Research Centre for Electronics and Telecommunication,
Indonesian Institute of Sciences, Indonesia,
Phone: 62 22 250 4661, Fax: 62 22 250 4659
Email: totok@ppet.lipi.go.id

ABSTRACT

In this report, it will be described the development of microsensor devices at the Research Center for Electronic and Telecommunication. The devices are microthermal conductivity detector and microtips for high electric field based gas sensor system. Microengineered thermal conductivity detectors (micro-TCDs) have been designed and fabricated on silicon using standard microelectronics processing. The device has an 100 x 100 mm^2 active area made from platinum/titanium metallization with a minimum linewidth of 10 mm. Each device has been diced and mounted on the header of a TO-5 package, and a unique detector cell was then designed as a means of its characterization. The micro-TCDs have shown fast responses to nitrogen gas under helium carrier. The intended application of such devices is in the field of gas chromatography. In this paper it will also be discussed the design and fabrication of a field ionization based gas sensor using micromachining technology. The core of this gas sensor is a microtip that has been fabricated on silicon using self aligned method. The process fabrication steps include the formation of a pyramidal silicon molding and electroplating. The fabrication of the silicon molding is done through wet chemical etching. With this technique, a microtip with tip edge radius of less than 4 mm has been successfully fabricated. It is expected that the microtip can be applied in the making of a field ionization based gas sensor.

Keywords: MicroTCD, Micromachining, Microtip, Gas Sensor, Field Ionization, Chemical Etching

Introduction

The main concern involved in the development of instruments for measuring gas parameters is to find a low cost gas sensor system with high sensitivity. With the advance in micromachining technology, however, it is now possible to design a low cost gas sensor with good sensitivity and selectivity. Two types of gas sensors that are of our interest are based on a field ionization method (Ghodsian *et al.*, 1998) and thermal conductivity. Field ionization method can be used to detect gases at standard atmospheric pressure and operated at low ionization voltage (Madau & Morrison 1991). The ionization currents generated in the ionization process can be used to detect species and concentration of the gas. The sensor has very low ionization current, in the order of nA so it requires a very low noise amplifier and a good wiring design. This time the research is focused on the fabrication process of field ionization tip with micromachining technology, and continued with the application of signal processing for data processing of ionization current generated by the sensor. The information obtained from ionization current using signal processing method is used to improve a sensor selectivity by means of a software. This is especially important if the gases to be detected are more than one species. The application of signal processing technique supported by the advantages of Personal Computer is possible to make a high performance gas instrument with low cost development.

On the other hand, TCD has been employed in many applications since the early days of gas chromatography (GC) development. It is one of the oldest and simplest types of sensors, which is known to have good stability and generality of response. Traditional TCDs consisted of two identical filaments made from temperature sensitive materials such as platinum, gold, nickel, or tungsten. In operation, the filaments were connected in a Wheatstone bridge configuration, with one filament exposed to a sample gas and the other filament exposed to a reference gas. Because of the difference in thermal conductivity of the sample and reference gases, signal variation resulted, detected by measuring the output voltage of the Wheatstone bridge. Further, the measured signal can be related to the concentration of the sample gas, allowing the sample gas to be identified (Hale *et al.*, 1992).

With advances in microengineering technology, TCD can be fabricated on planar substrates such as silicon, glass, or alumina. The filaments may be replaced by thin metal films made from the same materials as those used for traditional filaments and deposited on top of the substrates. Such TCD are commonly referred to as microengineered TCD (micro-TCD) to distinguish the fabrication technology. Thus, the fundamental principle of a micro-TCD and a traditional TCD are identical. The only significant difference is in the physical structures of the sensors. Being fabricated using microengineering technology, a micro-TCD gains additional advantages from the well-established microengineering infrastructures. These advantages include design flexibility, the ability to produce small devices, tight dimensional tolerances as determined by the lithography used, batch fabrication at low cost, reliability, reproducibility, high yield, easy fabrication of sensor arrays, and the possibility of producing control electronics on the same chip (Manginel *et al.*, 1997).

In this work, the chips of the micro-TCD consisted of metal tracks made from platinum/titanium. The active area has been designed in a meander pattern with minimum line-width of 10 µm. With the spacing between adjacent metal tracks of 12.5 µm, an active area as small as 100 µm x 100 µm has been achieved. The entire chip measures 1.5 mm x 1.5 mm. The chip has also been diced and mounted on a package to allow its characterization prior to connection to a GC system. In addition, a small measurement chamber (TCD cell) has been developed specifically for the purpose of performing the characterization of the micro-TCD. The TCD cell was made of Perspex material with hole located on its upper part as a gas entrance. The bottom part consists of cavity and exit holes, which permit immediate escape of the gas stream after contact with the chip. Such a cell structure minimizes the dead volume. Measurement of gas thermal conductivity can be performed in a flowing system as is common in GC situations.

The Micro Thermal Conductivity Detector

Design Consideration of the Detector Chip

There are several factors that should be taken into account when designing a micro-TCD for a GC system. The main one is related to the fact that a GC column must have had a finite volume such that a sample will be hard to detect without the detector having a fast thermal time constant and a high sensitivity. Therefore if the micro-TCD is dedicated as a GC detector, its geometry will be the most critical parameter. Secondly, there are some issues related to complexity of fabrication of a microchip, such as the use of thin insulating membrane underneath the active area (Van Herwaarden & Sarro 1988). These factors, along with other conditions such as the limit of photolithography, available materials and cost have defined to the necessary considerations during the design process.

The chips for the microengineered TCD have been designed with a structure as shown in Figure 4.1. It consists of a silicon substrate with two metal films deposited on its surface. The first (bottom) film served as an adhesion layer for the upper metal and the substrate, and so this layer is usually very thin (within a few hundred angstroms). The second (upper) metal formed an active layer, which will be exposed directly to gases. The same layer will also be used as the device bonding pads. Both metal films formed an active area and were electrically insulated from the substrate by a passivation layer of SiO_2. The major advantage of this structure is its simple fabrication process, which required only a single mask to define the metal patterns.

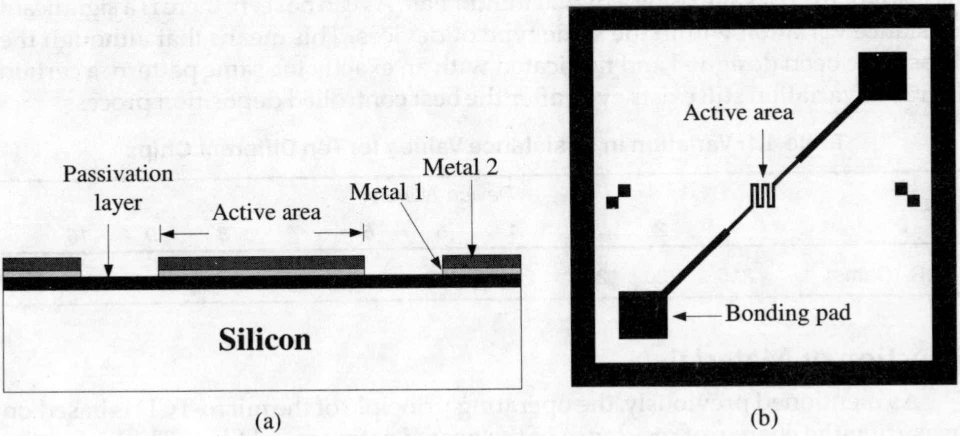

Figure 4.1: Structure of the Micro-TCD (a) Cross sectional view, and (b) Top view.

Since the microTCD is basically a resistive sensor with a response change based on changes in its resistance value, an important consideration in the design of the metal pattern is its nominal resistance. The total resistance of the active metal depends on its resistivity, film thickness, and length of the metal tracks. The principle of determining the resistance of a deposited metal film can be described as follows. In order to calculate the resistance of the metal film, it is required that the resistivity of the film structures be known. Very often this value is obtained by assuming that the deposited film has a continuous structure but, unfortunately, this is not the case with the metal films of the microengineered detector. There are three discontinuities that can be identified on the deposited film structures (Ciureanu & Middelhoek 1992). Firstly, there is a discontinuity within each metal film itself as a result of its deposition process. The fact that the growth of thin films is an additive process, in which the film is built up atom by atom, means that its structure will be discontinuous. The second discontinuity occurs at the surface between the first and second metal layers, and is due to the differences in the atomic properties of these layers. The third discontinuity occurs between the first metal layer and the dielectric layer underneath it. This discontinuity, in particular, must take into account the large ratio of the resistivities of metal and dielectric, which makes the electrical properties very sensitive to composition and structure of the film. Such discontinuities prevented the precise calculation of film resistivity of the deposited metal.

Therefore the resistance of the metal film of the microengineered TCD has been determined experimentally, and its exact value can be found by measuring the final resistance following photolithographic patterning of the film. The measured value takes into account the resistances of the bonding pads and the metals connecting the active area and the bonding pads. Table 4.1 shows the variations of the nominal resistance values as measured for several different chips. The chips were taken randomly from the wafer after scribing. The resistance was measured directly from the leads of the package using a digital multimeter. As can be seen, there is a significant resistance variation within the same type of devices. This means that although the chips have been designed and fabricated with an exactly the same pattern, a certain degree of variation still exists even after the best controlled deposition process.

Table 4.1: Variation in Resistance Values for Ten Different Chips

	Device Number									
	1	*2*	*3*	*4*	*5*	*6*	*7*	*8*	*9*	*10*
R (Ohms)	215	220	218	214	222	215	224	216	218	216

Selection of Materials

As mentioned previously, the operating principle of the micro-TCD is based on measuring the change of resistance of the heated active metal films. The higher the resistance change, the easier it is to measure using a basic electronic circuit. Thus an essential requirement for the metal films is that they should have a high resistivity and a high temperature coefficient of resistance (TCR). Table 4.2 shows the most common materials used for TCDs.

Table 4.2: Properties of Common Materials Used for TCDs

Metal	*Resistivity (μΩ cm)*	*TCR (ppm °C⁻¹)*	*Melting point (°C)*
Aluminium	2.69	4200	659
Chrome	19.0	5880	1903
Copper	1.67	4300	1083
Nickel	6.84	6844	1455
Gold	2.44	4000	1337
Platinum	10.3	3920	1770

Besides its resistivity and TCR, the metal film must have a good adherence with the substrate and be compatible with the externally bonded leads. The use of bonding pads from the same materials as the active film should greatly simplify the deposition steps and may require only a single mask. Another important issue with metal deposition on silicon is the ohmic character of the contact, caused by the difference in work functions of the metal and silicon (semiconductors in general). This difference leads to the bending of the semiconductor bands and results in either an accumulation or a depletion layer formed in the semiconductor. Furthermore, an accumulation layer results in ohmic contact and is preferred for electron transport.

For the developed TCDs, platinum (Pt) has been chosen as the active metal for its high resistivity. This metal can be deposited easily on silicon and bonded well with the external leads. Thus it eliminated the need for another metal such as aluminium for the bonding pads. Platinum has been known for its stability and reliability. Adhesion to the silicon substrate was enhanced using a thin layer of titanium (Ti).

Fabrication of the Micro-TCD Chips

The fabrication process sequence for the micro-TCD chips can be seen in Figure 4.2. It started with deposition of a 1.2-mm thick SiO_2 layer on the surface a 3-inch (100) silicon wafer by a wet oxidation process. The oxide layer will be used as an electrical insulation between the metal tracks and the silicon substrate. On top of the SiO_2 layer, a 300-Å Ti layer and a 1500-Å Pt layer were subsequently deposited. The deposition of these layers was performed using a sputtering process. These metals were then patterned using a standard positive-based photolithography process. Pattern transfer was done on a Canon (PLA-501FA) mask aligner with a bright field mask. Metal layers outside the sensor tracks and the bonding pads were then removed by a sputter etching process. The final step was to dice the individual chip and mount it onto the header of a TO-5 package. Figure 4.3b shows the resulting device after being diced and wire bonded onto the package.

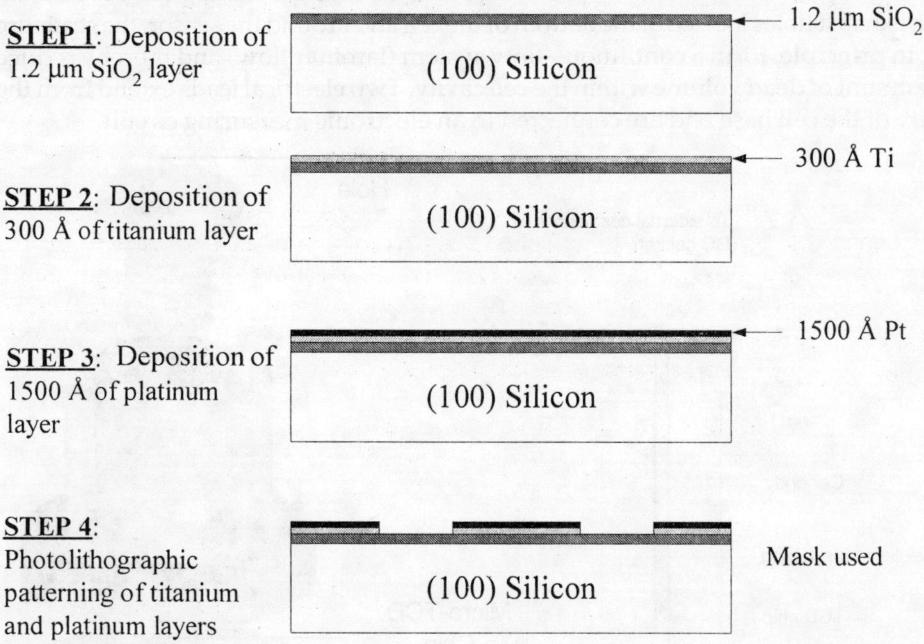

Figure 3.2: Fabrication Process Sequence for the Micro-TCD Chips

Design of the Detector Cell

The detector cell has several requirements to be met. In addition to being small in volume, the cell must be able to minimize any dead volume that can affect the sensor response during measurement. The resulting flow characteristics through the detector cell are also very important, especially for gas chromatography application, where these factors may lead to band broadening. Such effects can be reduced by using on-column detection, in which case the detector must be fabricated using the same process as the microengineered column. On-column detection techniques have been demonstrated by Manz *et al.* (1990) and Cowen & Craston (1995). Usually, to detect sample components eluting from a GC column, an amperometric-based measurement method is employed.

Figure 4.3 shows the schematic of the detector cell. The cell body has been fabricated from Perspex simply because of its transparency and relative ease to machine. The upper part of the cell has been designed to fit with a 1/32nd-inch outside diameter (OD) stainless steel tube. The tube itself has an inside diameter (ID) of 100 mm and it can be used to transport a sample gas from an external gas source or from a GC column. The detector chip, mounted on its package, is located at the bottom of the cell (within the cell cavity) with its active area positioned in the centre of the flow path and just below the stainless steel tube hole. Such positioning allows an instantaneous contact between the detector chip and the incoming gas, so that immediate gas detection may be performed and time for gas mixing can be reduced. With exhaust holes located at the bottom of the cell and around the sensor, the structure can, in principle, form a continuous-flow system (laminar flow) and thereby reduce the amount of dead volume within the cell cavity. Two electrical leads extend from the centre of the cell base and are connected to an electronic measuring circuit.

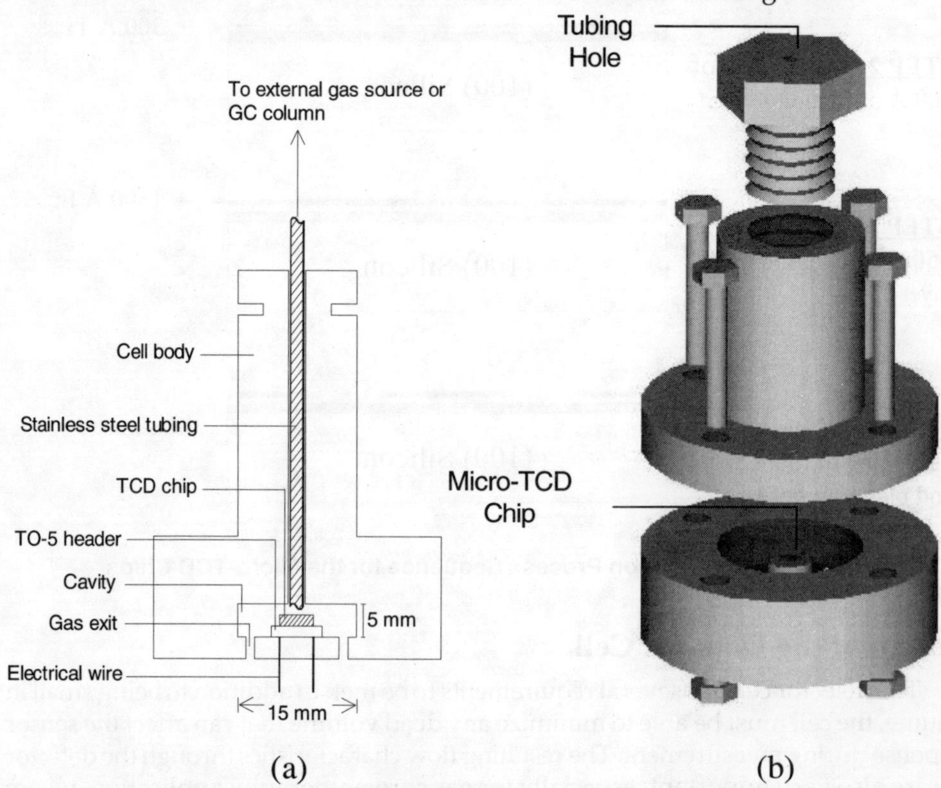

Figure 4.3: Schematic of the Detector Cell
(a) Two dimensional view, and (b) Three dimensional representation

Performances of the Micro-TCD Chips

The micro-TCD chips have been evaluated for their responses to gases. The detector chip is located inside the detector cell, and gases then flow to the surface of the detector chip by means of a stainless steel tube. To simulate the actual geometry of a GC column, the internal diameter (ID) of the stainless steel tube has been chosen to be 100 mm, which is typical for modern GC columns. Due to such a small diameter, gases entering the tube will experience a large pressure differential, so that the resulting gas linear velocity and flow rate at the tube exit will also be small. To be able to detect incoming gases with a small flow rate, the detector chip has been positioned very close to the tube exit. Such positioning is possible with this particular cell, which has been designed to achieve an approximately 500-mm gap between the top surface of the chip and the tube exit hole.

Two different gases may be supplied simultaneously. A miniature union has been used to regulate the two gases directly from their cylinders. With this union, gases coming from each cylinder are mixed, and directed towards the 100-mm tube connected to the detector cell. With this set-up, the detector chips can be evaluated for their responses to binary gas mixtures. The detector will then respond to one gas with the background of the other. Alternatively, one of the gas cylinders may be turned off to allow a single gas to pass through the detector. In such cases, the detector will respond to the gas under the background of air.

Figure 4.4 shows a typical response of the detector with the introduction of nitrogen gas under still air. The measurements were taken under three different inlet pressures: 20, 40, and 60 psi. The plots obtained for each pressure value have been superimposed for comparison purposes. Also, the vertical axis indicates the net change in the output voltage relative to its initial steady state value. In other words, the zero value on the vertical axis represents the steady state condition under the influence of the supply current only. With an initial detector temperature of 175 ºC, it takes less than 1 minute to reach the second steady state condition. To further evaluate the performance of the detector, a series of injections has been performed on the system. As before, nitrogen was injected under helium carrier gas, but the duration of each injection was only 1 second. Once the response reached the original baseline, a second injection was performed and this cycle could be repeated. The importance of this procedure is to simulate the response of the detector to successive peaks from a GC column with the duration of 1 second for each peak (the actual GC peaks may have durations of a few seconds). The result of three such successive injections is shown in Figure 4.6. As can be seen, the detector still showed a fast response even under this condition. Despite a slightly longer recovery time, the rise time for each pulse is very short.

The Field Ionization Detector

The principle of field ionization detection can be described as follows. Two parallel plate electrodes with the applied voltage Vo initially have a uniform electric field. An uncharged gas molecule between these plates will experience polarization and become dipole molecules. The positive and negative dipoles experience a Coulomb

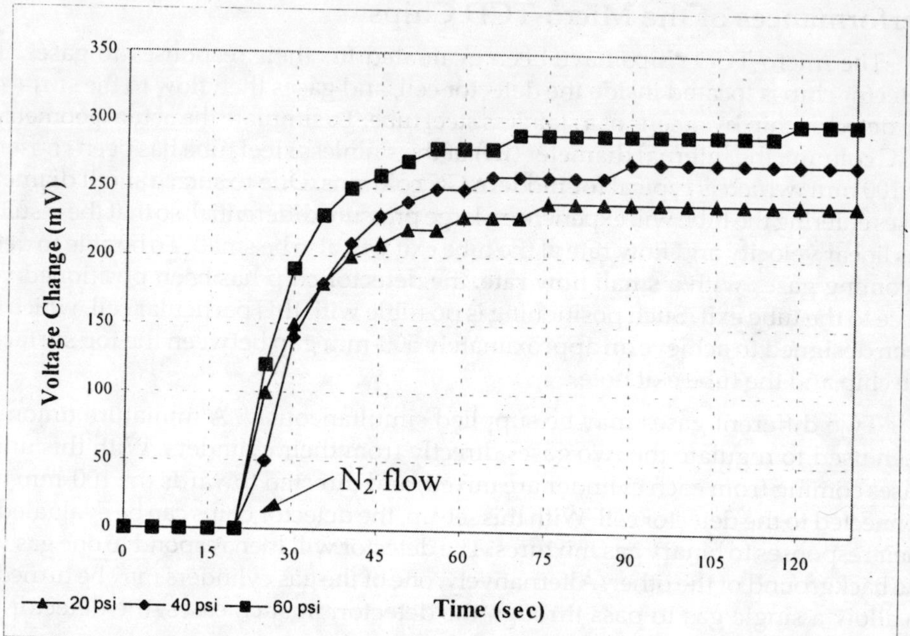

Figure 4.4: Detector Response to Nitrogen Gas in Still Air Under Various Tube Inlet Pressures: First Steady State Temperature = 175°C

force by the electrodes with the same force but have opposite direction so the resultant is zero. This condition will go on as long as the electric voltage between the electrodes is less than ionization voltage of the gas molecule.

If the electric voltage between the electrodes is unchanged, but the electric field distributions between the electrodes become nonuniform by changing positive electrode with a sharp edge electrode (tip), then a negative edge of a dipole molecule will experience more force than a positive edge. The negative dipole will be attracted to the positive electrode. If the electric field around the tip is greater than electric field for ionizing the gas molecule, then the electron of the gas molecule will be attracted to the positive electrode. The gases then become a positive gas ion and will move toward a negative electrode and at the outside circuit ionization current flows.

The electric fields around the tip have a high intensity field that depends on the radius of the tip edge and the voltage of the tip. The value of the electric field around the tip can be calculated with the equation:

$$\mathbf{E = V_o/kr} \qquad \textbf{(Volt/m)} \qquad (1)$$

where,

V_o = electric field at electrode tip, r = radius of the tip edge, and k = constant.

Gas ionization happened at a sharp edge, fine wire, or anything that can generate electric field greater than electric field to breakdown the gas. The electric field to

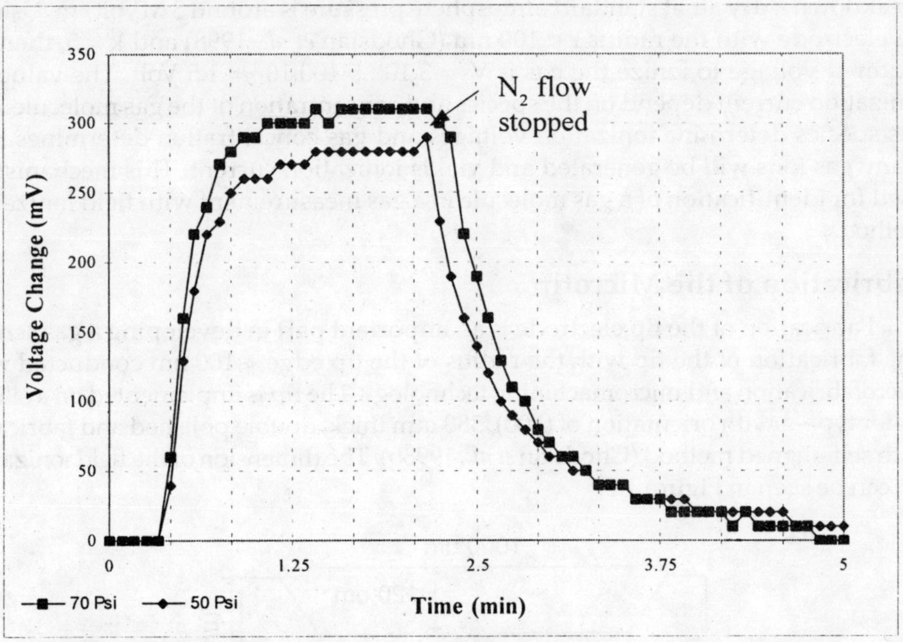

Figure 4.5: Recovery Time of the Detector Under Nitrogen Gas in Still Air for Two Different Inlet Pressures: First Steady State Temperature = 175 °C

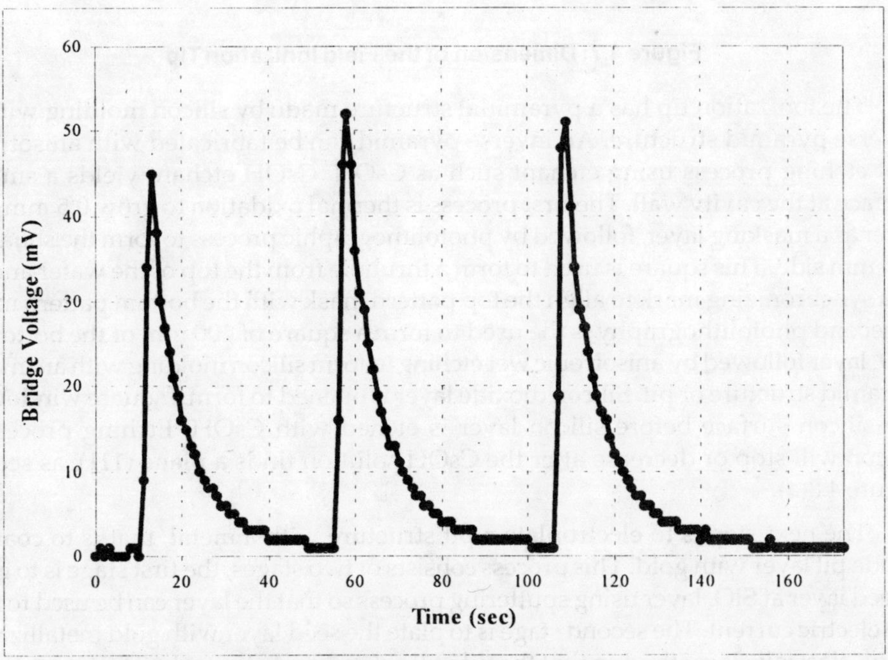

Figure 4.6: Response of the Detector to Three 1-second Injections of Nitrogen Under Helium Carrier Gas

breakdown a dry air at standard atmosphere pressure is around 3 MVolt/m. Using a tip electrode with the radius r < 100 nm (Ghodsian *et al.*, 1998) and k = 5, then the required voltage to ionize the gas is $V_o = 3.10^{+6}.5.100.10^{-9} = 1.5$ Volt. The values of ionization current depend on the species and concentration of the gas molecule. The gas species determine ionization voltages and gas concentration determines how many gas ions will be generated and yields ionization current. This mechanism is used for identification of a gas molecule in a gas measurement with field ionization method.

Fabrication of the Microtip

Fabrication of the tip electrode is an important part in developing a gas sensor. The fabrication of the tip with the radius of the tip edge < 100 nm conducted with microfabrication and micromachining technology. The tip is implemented on a silicon wafer type-n with orientation of (100), 380 mm thick, double polished and fabricated with self-aligned method (Ghodsian *et al.*, 1998b). The dimension of the field ionization tip can be seen in Figure 4.7.

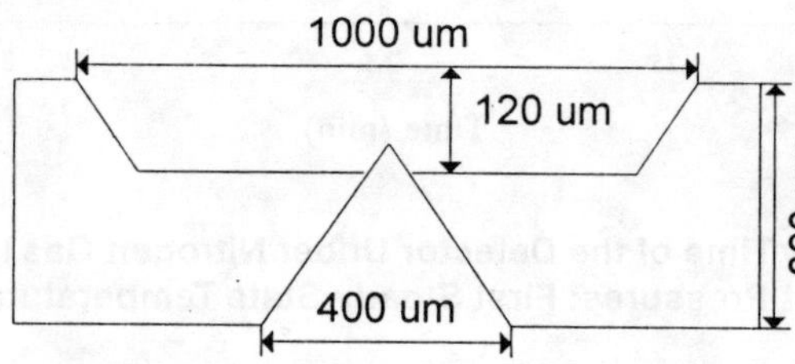

Figure 4.7: Dimension of the Field Ionization Tip

The ionization tip has a pyramidal structure made by silicon molding with an inverse pyramid structure. An inverse pyramid can be fabricated with anisotropic wet etching process using etchant such as CsOH. CsOH etchant yields a smooth surface at the cavity wall. The first process is thermal oxidation to grow 0.5 mm SiO_2 layer as a masking layer, followed by photolithographic process to form the square of 700 mm side. This square is used to form a thruhole from the top of the wafer and act as a synchronizing mark to align the top pattern mask with the bottom pattern mask. A second photolithography is the used to form a square of 400 mm of the bottom of SiO_2 layer followed by anisotropic wet etching to form silicon molding with an inverse pyramid structure or pit. Silicon dioxide layer is opened to form a square window on the silicon surface before silicon layer is etched with CsOH. Etching process of silicon will stop or decrease after the CsOH solution finds a plane (111), as seen in Figure 4.8(a).

The next step is to electroplate a pit structure with a metal, that is to coat the inside pit layer with gold. This process consists of two stages, the first stage is to grow a seed layer at SiO_2 layer using sputtering process so that the layer can be used to flow the electric current. The second stage is to plate the seed layer with gold metallization using electrolysis method so that the pit will act as a pad electrode. The last step is to eliminate the rest of the SiO_2 layer and the sensor is ready for packaging, see Figure

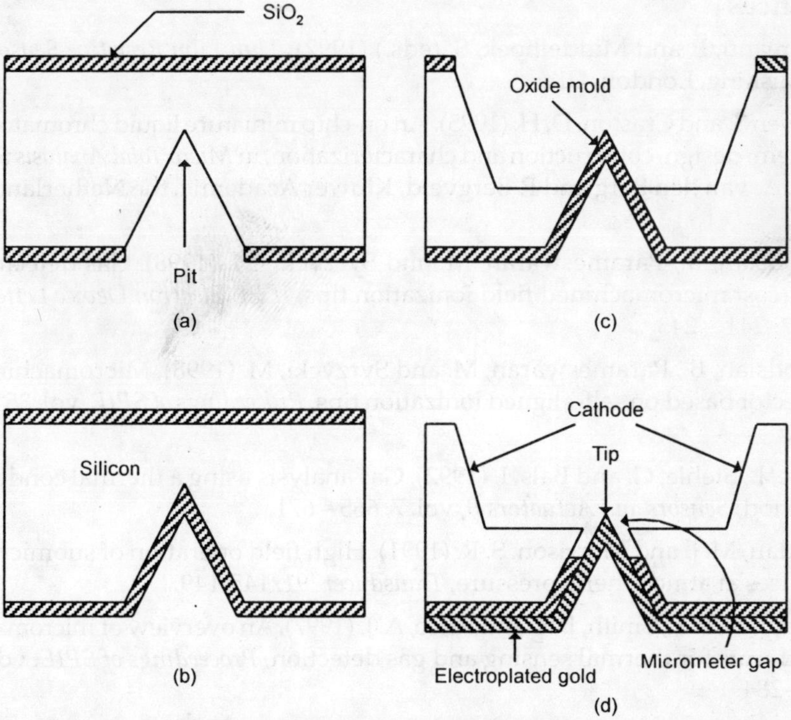

Figure 4.8: The Fabrication Sequence of One Single Tip

4.8 (d). To investigate the effect of the number of the tips to selectivity, in this research it has designed various topologies of the ionization tips, including a single tip, a double tips or an array of tips.

Conclusion

The design and fabrication of two micro devices have been described in this paper. The first part is the microTCD, which was fabricated on silicon having an active area of 100 x 100 mm^2. Platinum has been used as the sensing element with a minimum linewidth of 10 mm. Each device has been mounted on the header of a TO-5 package and placed inside a unique detector cell, where its characterization was performed. Each of the detector chips tested was able to show fast responses to nitrogen gas under still air and helium background. An excellent response of the detector to GC peaks, simulated by successively injecting three 1-second pulses of nitrogen gas under helium, has also been demonstrated.

The second part of this paper describes the concepts of field ionization tip fabricated with micromachining technology. It has been demonstrated that the self-aligned method was capable of producing microtips having a tip edge radius of less than 4 mm. The process fabrication steps include the formation of a pyramidal silicon molding and electroplating. The fabrication of the silicon molding is done through wet chemical etching. It is expected that the microtip can be applied in the making of a field ionization based gas sensor.

References

- Ciureanu, P. and Middelhoek, S. (eds.) (1992). *Thin Film Resistive Sensors*, IOP Publishing, London, UK.
- Cowen S. and Craston, D. H. (1995). An on-chip miniature liquid chromatography system: design, construction and characterization, in *Micro Total Analysis Systems*, eds. A. van den Berg and P. Bergveld, Kluwer Academic, the Netherlands, 295-298.
- Ghodsian, B., Parameswaran, M. and Syrzycki, M. (1998). Gas detector with low-cost micromachined field ionization tips, *IEEE Electron Device Letters*, vol. 19, 7: 241 – 243.
- Ghodsian, B., Parameswaran, M. and Syrzycki, M. (1998). Micromachined gas detector based on self-aligned ionization tips, *Proceedings of SPIE*, vol. 3514, 403-410.
- Hale, J., Stehle, G. and Bals, I. (1992). Gas analysis using a thermal conductivity method, *Sensors and Actuators B*, vol. 7, 665 – 671.
- Madau, M. J. and Morrison, S. R. (1991). High field operation of submicrometer devices at atmospheric pressure, *Transducer '91*, 145-149.
- Manginel, R. P., Smith, J. H. and Ricco, A. J. (1997). An overview of micromachined platforms for thermal sensing and gas detection, *Proceedings of SPIE*, vol. 3046, 273-284.
- Manz, A., Miyahara, Y., Miura, J., Watanabe, Y., Miyagi, H. and Sato, K. (1990). Design of an open-tubular column liquid chromatograph using silicon chip technology, *Sensors and Actuators B1*, 249-255.
- Van Herwaarden, A. W. and Sarro, P. M. (1988). Floating-membrane thermal vacuum sensor, *Sensors and Actuators*, vol. 14, 259-268.

5

Implant Fabrication of Single Crystal Silicon MEMS: A Novel Processing Route

Ahmed Shuja
Associate Professor,
COMSATS Institute of Information Technology, Islamabad, Pakistan
E-mail: shuja_ahmed@comsats.edu.pk

ABSTRACT

The development of a new technology for integration of dissimilar materials will be needed to meet many of the advanced technological needs of the future. Current technologies are limited by the ability to integrate materials systems together on a single platform (*e.g.*, a Si wafer for applications in microelectronic, optoelectronic, and microelectromechanical systems – MEMS). A layer splitting process by ion-implantation and subsequent annealing, called 'ion-cut', is an innovative approach for engineering materials that provides an effective, versatile, and economical method of transferring thin surface layers from bulk substrates onto a host of other substrates. This research presents a demonstration of a single crystal silicon MEMS fabrication technology using hot proton-implantation smart-cut technique. The proposed technology, compared to conventional SOI wafer fabrication processes for MEMS applications, can potentially result in a significant substrate and processing cost reduction. Prototype cantilever beams and clamped-clamped micro-bridges have been successfully fabricated as demonstration vehicles for future microsystem implementations. The process is also studied by elastic recoil detection (ERDA) and Rutherford backscattering in channelling mode (RBS) to obtain profiles of the implanted hydrogen and the damage. An end product for the prototype is investigated by Scanning Electron Micrography (SEM).

Keywords: MEMS, Ion Implantation, Smart cut, Hot Implantation, Single Crystal Silicon.

Introduction

The continual demand for sensors and control systems in all areas has stimulated the development of micro-electro-mechanical systems (MEMS). For this purpose, many alternatives of materials and fabrication processes are applied and new alternatives are frequently experienced. In general, integration between mechanical microstructures and microelectronic circuits is preferred due to advantages offered by the well established microelectronic technology, such as batch process, miniaturization, low cost, fast response, low noise, and the possibility to realize in situ signal processing. Moreover, silicon has appropriated mechanical and electrical characteristics for microstructures fabrication. In such way, it is important that materials and processes for fabrication of integrated microstructures be compatible with CMOS technology.

The development of a new technology for integration of dissimilar materials will be needed to meet many of the advanced technological needs of the future. Current technologies are limited by the ability to integrate materials systems together on a single platform (*e.g.*, a Si wafer for applications in microelectronic, optoelectronic, and microelectromechanical systems – MEMS). A layer splitting process by ion-implantation and subsequent annealing, called 'ion-cut', is an innovative approach for engineering materials that provides an effective, versatile, and economical method of transferring thin surface layers from bulk substrates onto a host of other substrates. In the ion-cut process, hydrogen is implanted into the material. Subsequently, the implanted wafer is bonded to a 'stiffener', here a silicon wafer capped with a silicon oxide layer. In the final step the bonded wafers undergo thermal annealing at elevated temperatures which leads to the formation of H-filled gas bubbles in the implanted wafer. These gas bubbles grow in size via Ostwald ripening, and ultimately provide the force to induce cleavage in the hydrogen-implanted water.

One application of ion-cut is the production of silicon-on-insulator (SOI) materials by the smart-cut™ process. While research has been conducted to achieve knowledge about how hydrogen interacts with silicon at the atomic level, a fundamental understanding of the basic mechanism on how the cutting process occurs is still unclear.

The Smart-Cut® process is based on ion implantation and wafer bonding, as detailed in Figure 5.1a. Ion implantation (hydrogen, helium) induces formation of an in-depth weakened layer located close to the mean ion penetration depth which leads to delamination of a thin film from a thick substrate. Wafer bonding enables different materials to be associated to form multi-layer substrates. As it allows many

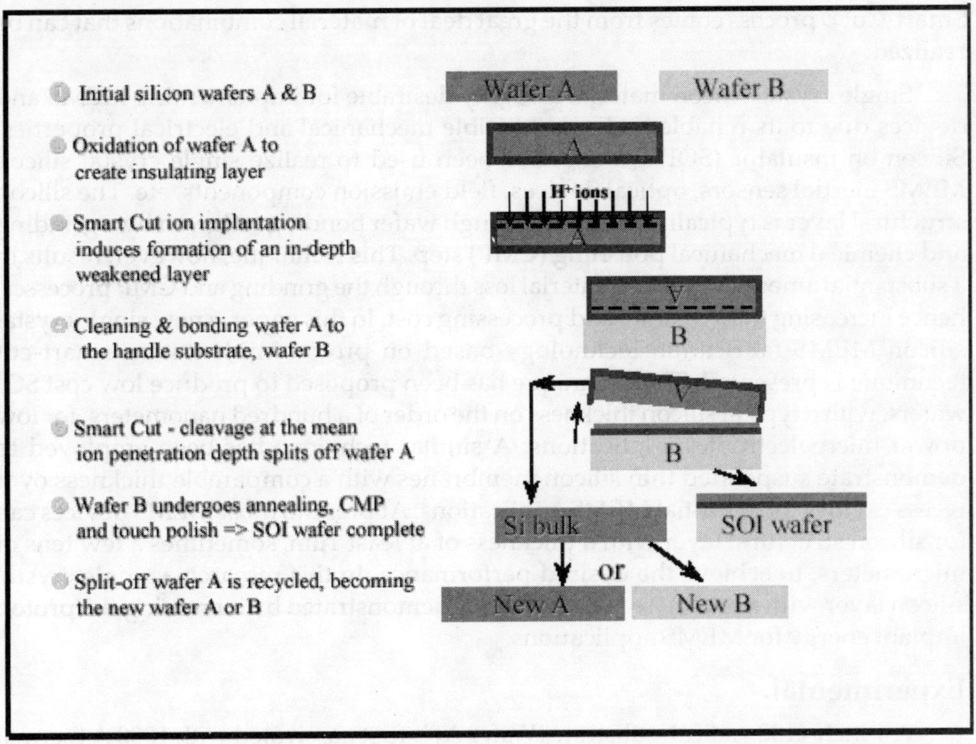

Figure 5.1a: Process Flow-Schematic of a Standard Smart-Cut Routine

material combinations, a lot of studies are under way to develop new structures using this process.

This process has several advantages from both technical and economic points of view. Among these, the possibility of transferring thin layers of different thickness with very good thickness homogeneity and with a high crystalline quality onto different materials deserves to be highlighted. From an economic point of view, the possibility of reusing the remainder of the implanted substrate and the process compatibility with standard microelectronic equipments help in reducing costs.

The most advanced studies on this process have been carried out on silicon. These studies have put in evidence physical mechanisms involved in this process and the relation between experimental conditions and splitting. Today, this process is industrially used to obtain Silicon On Insulator wafer. The flexibility of the process in regards of the thickness of the top silicon layer and buried oxide layer allows the formation of different SOI structures. Silicon on glass or on fused silica is other kind of SOI wafers, which can be obtained thanks to this process. Thus, we demonstrated that the Smart-Cut® process can lead to transfer of a thin film and reclaiming of the seed substrate even for structures where the two materials exhibit different thermal expansion coefficients such as silicon and fused silica. The generic nature of the

Smart-Cut® process comes from the great deal of material combinations that can be realized.

Single crystal silicon material is highly desirable for implementing MEMS and devices due to its reliable and reproducible mechanical and electrical properties. Silicon on insulator (SOI) wafers have been used to realize single crystal silicon MEMS inertial sensors, optical devices, field emission components, etc. The silicon structural layer is typically obtained through wafer bonding followed by a grinding and chemical mechanical polishing (CMP) step. This technique, however, results in a substantial amount of silicon material loss through the grinding and CMP processes; hence increasing the substrate and processing cost. In this paper, a new single crystal silicon MEMS fabrication technology based on proton-implantation smart-cut technique is presented. This technique has been proposed to produce low cost SOI wafers, with a typical silicon thickness on the order of a hundred nanometers, for low power microelectronics applications. A similar technique has been employed to demonstrate suspended thin silicon membranes with a comparable thickness over recess cavities for potential MEMS applications. At present, most MEMS devices call for silicon structural layer with a thickness of at least 1μm, sometimes a few tens of micrometers, to achieve the desired performance. In this research a single crystal silicon layer with a thickness close to 2μm is demonstrated by increasing the proton implant energy for MEMS applications.

Experimental

A 4-inch <100> silicon substrate (Wafer A) is first passivated with 1000 Å thermal oxide followed by the 185 keV proton Implantation, to a fluence ranging from 2 x 1016 to 9 x 1016 ions/cm2, at sample temperatures of "90 °C, 30 °C, 100 °C, 150 OC, 200 °C and 300 °C respectively. The implantation energy is chosen to achieve a H+ peak concentration of approximately 1.75 μm below the wafer surface. Each implant was studied by elastic recoil detection (ERDA) and Rutherford backscattering in channelling mode (RBS) to obtain profiles of the implanted hydrogen and the damage. Changes in the strain profile of the implant were monitored using double axis X-ray diffraction. The wafer temperature was measured by a thermocouple attached to a silicon guard ring which held the sample in place. The temperature of the silicon guard ring was calibrated against the temperature of the sample. Samples from each of the wafers were inspected by SEM high field atomic force microscopy (AFM) mode and Nomarski optical microscopy before and after annealing to investigate blister formation. Some samples were also studied by transmission electron microscopy.

Process Characterization, Results and Discussion

Figure 5.1b shows an evolution of blister density and mean blister diameter as deduced from the channelling RBS spectra and corresponding ERD for the devices processed by the hot implanted smart cut procedure. The lattice disorder of the samples, proportional to the normalised counts on these spectra, grows with increasing temperature between 30°C and 200°C. Transmission electron microscopy images of the 200°C sample show that the hydrogen agglomerates into the platelets, microcracks and bubbles which are responsible for blistering and exfoliation, in agreement with

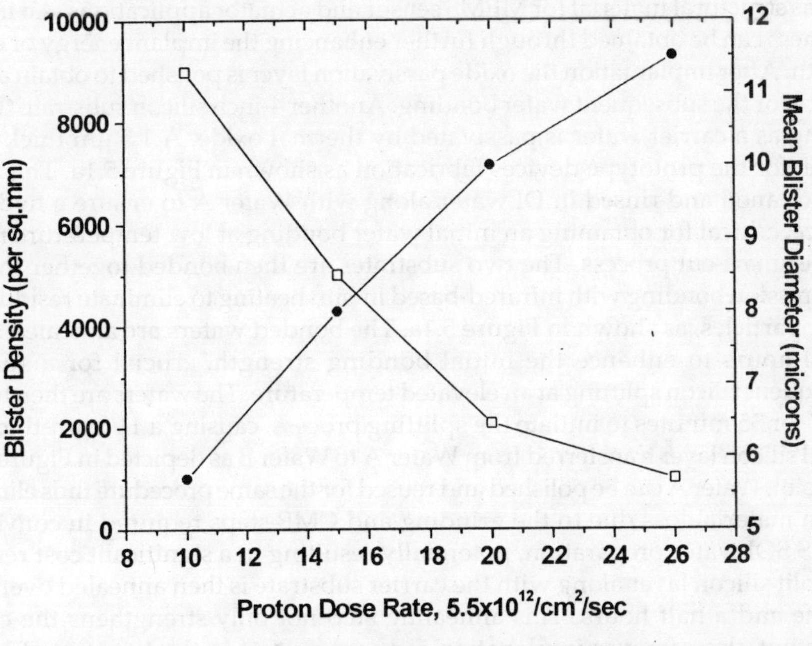

Figure 5.1b: An Evolution of Blister Density and Mean Blister Diameter as Reduced from the Channelling RBS Spectra and Corresponding ERD for the Devices Processed by the Hot Implanted Smart Cut Procedure

previous studies. Therefore we assign the increase in the damage to the rearrangement of point defects into these defect complexes. As the implant temperature is raised, the strain is relieved, indicated by the attenuation of fringes to the left of the high intensity substrate peak. This is in agreement with the idea that at higher implant temperatures, hydrogen becomes more mobile in the lattice, and forms into complexes, such as platelets, which relieve the strain.

Qualitatively it is known that implanted hydrogen evolves under heating to form bubbles with high internal pressure which then drives the cleavage process. The general belief has been that the bubbles which induce the cleavage for the ion-cut occur at the peak in the H-implantation concentration profile. However, our recent results have led us to speculate that H bubble nucleation and ultimate cleavage are controlled by the lattice damage that is generated by the H-implantation process. Both theory and past experiment have shown that the damage profile resulting from ion-implantation has a mean depth that is shorter than the ion projected range. In this work, we show that the cut depth is controlled by the damage depth rather than the implantation projected range.

The implanted hydrogen ions introduce micro-cavities along the peak concentration. After bonded to a carrier substrate, the micro-cavities cause silicon wafer to split along the peak concentration when annealed at elevated temperatures, thus forming a uniform crystal silicon layer with a thickness of 1.78 μm, which can be used as structural material for MEMS sensor and actuator applications. An increased thickness can be obtained through further enhancing the implant energy or epitaxial growth. After implantation the oxide passivation layer is polished to obtain a smooth surface for the subsequent wafer bonding. Another 4-inch silicon substrate (Wafer B) serving as a carrier wafer is passivated by thermal oxide. A 1.5 μm thick oxide is chosen for the prototype devices fabrication as shown in Figure 5.1a. This wafer is RCA cleaned and rinsed in DI water along with Wafer A to ensure a hydrophilic surface, critical for obtaining an initial wafer bonding at low temperature required for the smart-cut process. The two substrates are then bonded together through a compression bonding with infrared-based in-situ heating to eliminate residual water vapor particles, as shown in Figure 5.1a. The bonded wafers are annealed at 250ºC for 10 hours to enhance the initial bonding strength, crucial for a successful subsequent silicon splitting at an elevated temperature. The wafers are then heated at 510ºC for 35 minutes to initiate the splitting process, causing a 1.78 μm-thick single crystal silicon layer transferred from Wafer A to Wafer B as depicted in Figure 5.1a. At this point, Wafer A can be polished and reused for the same procedure thus eliminating silicon material loss due to the grinding and CMP steps required in conventional MEMS SOI wafer preparation, potentially resulting in a significant cost reduction. The split silicon layer along with the carrier substrate is then annealed over 1050 °C for one and a half hours. This annealing step not only strengthens the chemical bonds but also removes implanation-induced defects in the transferred layer. To experiment the prototype process, sample pieces diced from Wafer A are bonded to Wafer B for the splitting study. Figure 5.2 presents an SEM photo of the wafer cross-section after completing the process flow described above, indicating that a 1.78 μm-thick single crystal silicon layer is transferred onto the carrier wafer, which is passivated by a 4000 Å-thick thermal oxide in this case.

A corresponding SEM photo of the sample surface is presented in Figure 5.3, showing a smooth surface of a part of the split silicon region over the carrier substrate. The void within the silicon region corresponds to an unbonded area due to trapped particles during bonding. Therefore, wafer surface cleanliness is crucial for a successful splitting. At this point, the silicon layer is patterned with standard

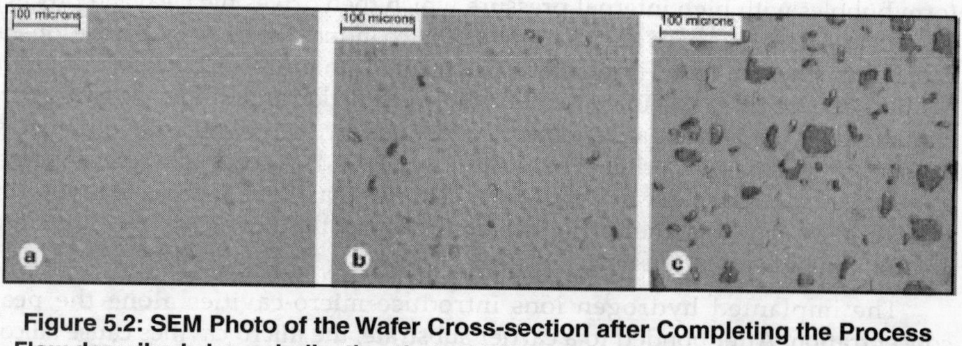

Figure 5.2: SEM Photo of the Wafer Cross-section after Completing the Process Flow described above, indicating that a 1.78 μm-thick Single Crystal Silicon Layer is Transferred onto the Carrier Wafer, which is Passivated by a 4000 Å-thick Thermal Oxide in this Case

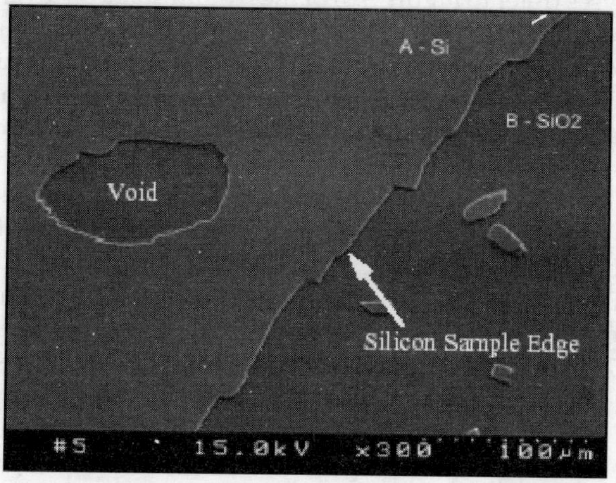

Figure 5.3: AFM Mode (SEM) Monograph of a Split Si MEMS Prototype Sample Surface

lithography procedure to form various microstructures such as cantilever beams, clamped-clamped micro-bridges, etc. A number of prototype microstructures have been successfully fabricated using the proposed fabrication technology. Figure 5.4 presents SEM photo of fabricated cantilever beams and clamped-clamped micro-bridges. These microstructures exhibit a width of 4 μm and lengths ranging from 90 μm to 187 μm. It can be seen that the beams are free released with an air gap of approximately 1.5 μm. The tip of the longest cantilever beam resides on the substrate possibly caused by stiction. An SEM inspection also reveals that the released cantilever beams are virtually free of warpage, thus corresponding to a negligible film strain gradient, a key advantage for MEMS applications. These demonstrated structures will be used as building blocks to design and fabricate more complex MEMS devices such as pressure, inertial, bio-chemical sensors, etc., planned as the next step.

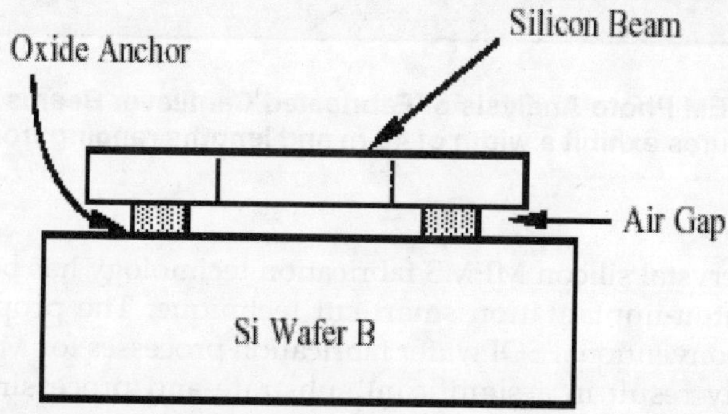

Figure 5.4: End Product of the Proposed Single Crystal Silicon MEMS Fabrication Process Flow: A released microstructure is shown whereas a proton fluence ranging from $2x10^{16}$ to $9x10^{16}$ cm^{-2} at 150°C is found adequate for silicon layer splitting.

TEM analyses (refer to Figure 5.5) of the resulting silicon layer reveal a negligible implant-induced defect density due to high temperature annealing and single crystal material characteristics, which are highly desirable for potential integration of MEMS devices with sensing and control electronics in the same structural layer for building high-performance micro-systems. Furthermore, the silicon film thickness can be accurately determined through an implant energy control, which presents a key advantage for high-precision devices fabrication. The proposed technique eliminates the grinding and CMP processes required in conventional MEMS SOI wafer preparation, potentially resulting in a significant substrate and processing cost reduction. MEMS prototype structures, such as cantilever beams and clamped–clamped micro-bridges with a typical length of a few hundreds of micrometers and a thickness of 1.7μm, have been fabricated as demonstrated vehicles for future sensors, actuators, and micro systems implementation. A thicker silicon layer can be obtained by enhancing the implant energy but will be ultimately limited by the equipment capability. An additional epitaxial growth on top of the existing silicon layer can be potentially considered to increase the layer thickness.

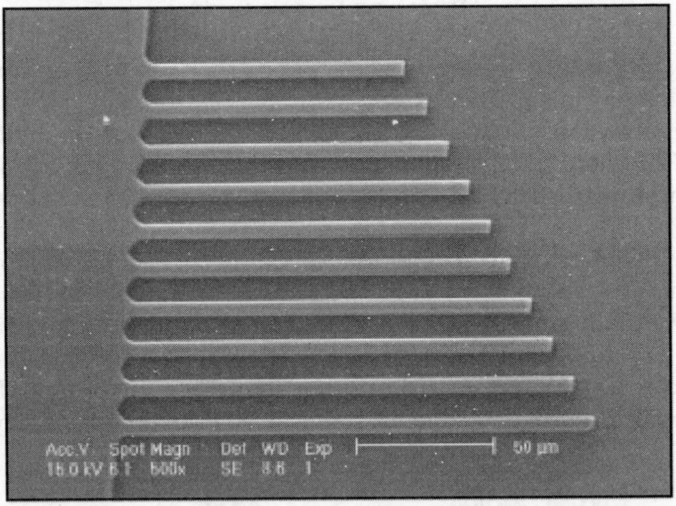

Figure 5.5: TEM Photo Analysis of Fabricated Cantilever Beams Top View: These microstructures exhibit a width of 4μ m and lengths ranging from 90 to 187 μm.

Conclusion

A single crystal silicon MEMS fabrication technology has been demonstrated using hot proton-implantation smart-cut technique. The proposed technology, compared to conventional SOI wafer fabrication processes for MEMS applications, can potentially result in a significant substrate and processing cost reduction. Prototype cantilever beams and clamped-clamped micro-bridges have been successfully fabricated as demonstration vehicles for future microsystem implementations. The proposed technology is also attractive for realizing other single crystal thin films such as SiC, GaAs, GaN, etc. for implementing MEMS sensors and electronic devices for harsh environment applications and potential integration of micro-systems with optoelectronics.

Acknowledgments

This work is partially supported by NSF under Bell labs innovation partnership programme. Fabrication process was performed at Micro-Fabrication Laboratory at Case Western Reserve University and University of Surrey, UK.

References

- Ahmed, S., Amirov, K., Larsson, U., Lin, J., Haq, A., Too, P., Tabatabaian, Z., 2005, VACUUM - Surface Engineering, Surface Instrumentation & Vacuum Technology, vol. 78, pp. 137-141
- Brosnihan, T., Bustillo, J., Pisano, A., Howe, R., 1997, Embedded Interconnect and Electrical Isolation for High-Aspect-Ratio, SOI Inertial Instruments, *International Conference on Solid-State Sensor and Actuators*, pp. 637-640.
- Bruel, M., 1995, *Electron. Lett.*, **31** (14), p. 1201.
- Conant, R., Nee, J., Lau, K., Muller, R., 2000, A Flat High-Frequency Scanning Micromirror, *IEEE Solid-State Sensor and Actuator Workshop*, pp. 6-9.
- Du, J., Ko, W., Young, D., 2004, Single crystal silicon MEMS fabrication technology Du, W. Ko, D. Young, Sensors and Actuators A 112, pp. 116-121
- Freund. L., 1997, *Appl. Phys. Lett.* **70**, p. 3519.
- Milanovic, V., Doherty, L., Teasdale, D., Parsa, S., Zhang, C., Pister, K., 2001, Micromachining Technology for Lateral Field Emission Devices, *IEEE Transaction on Electron Devices*-special issue on vacuum electronics, Vol. 48, No. 1, pp. 166-173.
- Milanovic, V., Last, M., Pister, K., 2001, Monolithic Silicon Micromirrors with Large Scanning Angle, *Optical MEMS 2001*, Okinawa, Japan, pp. 135-136.
- Parameswaran, L., Hsu, C., Schmidt, M., 1995, A Merged MEMS-CMOS Process Using Silicon Wafer Bonding," *IEDM*, pp. 613-616.
- Sangwoo Pae, Taichi Su, John P. Denton, and G. W. Neudeck, 1999, Multiple Layers of Silicon-on-Insulator Islands Fabrication By Selective Epitaxial Growth," IEEE Electronic Device Letters, Vol. 20, No. 5, pp. 194-196.
- Tong, Q.-Y., Gutijahr, K., Hopfe, S., and Lee, T.-H., 1997, Layer Splitting Process in Hydrogen-Implanted Si, Ge, SiC, and Diamond Substrates," Applied Physics Letters, 70. pp. 1390-1392.
- Varma, C., 1997, Appl. Phys. Lett. 71, p. 3519.

6

Soft Lithographic Techniques and their Applications in Rapid Fabrication of Micro and Nano Structures in Polymers and Metals

S.C. Jain
National Physical Laboratory,
Dr. K. S. Krishnan Road, New Delhi – 110 012
E-mail: scjain@mail.nplindia.ernet.in

ABSTRACT

Rapid fabrication of small structures with dimensions in the range of a few tens of microns to a few hundred nanometers is central to the development of miniaturized systems and devices that include microelectronics, sensors, microreactors, total microanalytical systems (mTAS), micromechanical systems (MEMS) and micro-optical systems. Consequently, there have been very fascinating developments in the field of fabrication of small structures in recent years.

We wish to describe below a new set of novel non-photolithographic techniques (soft lithographic techniques) to fabricate micro and nano structures in various polymers and metals. Primarily, a desired pattern structure is created on a silicon/glass substrate using standard photolithography and etching techniques. The prepolymer of poly dimethyl siloxane (PDMS) is cast against the master pattern structure, cured and peeled from the master to create a master polymer stamp. The cured polymer has the features that are replica of the master structure. The master stamp thus prepared exhibits high fidelity and can be used in a variety of ways to produce small structures in different polymers and metals. The different processes have been classified as: micro contact printing (mCP), micro transfer molding (mTM), micro replica molding (mREM), micro molding in capillary (MIMIC), and solvent assisted micromolding (SAMIM). These techniques have been used to produce patterned structures in various coinage metals like silver and gold, poly methyl metha acrylate (PMMA), polyurethane (NOA-65), and replication of optical grating. The techniques are cost effective, experimentally convenient and are being actively explored at several places. They hold a great promise for fabricating microelectrodes, microfluidic networks and devices, biosensors, mTAS etc.

Introduction

The field of nanoscience and technology has witnessed a phenomenal growth in last one decade. It has been identified as one of the priority areas of future materials science & technology world over and huge sums of money has been pumped into this research field by various government and private funding agencies. Being a multidisciplinary research field, it offers many exciting and challenging opportunities to materials scientists, chemists, physicists, biologists and engineers alike. It has been recognized that the today's R&D efforts into areas such as molecular self-assembly, novel patterning techniques, precision metrology tools, nano-structure materials and micro mechanical systems would form the building blocks of tomorrow's material science and engineering.

Fabrication of functional structures with different shapes and sizes in various metals and polymers is central to the development of many technologies involving nano materials and structures. It is also desirable to create surfaces that exhibit patterns with contrasting properties such as wetability, chemical reactivity and surface charges down to micron and submicron level on various substrates for applications in microelectronics, optoelectronics, MEMS, microfluidic networks and devices, sensors & arrays, total micro analytical systems (mTAS) etc. Microfabrication coupled with automation science is going to result in the development of new diagnostic tools for pharmaceutical and biotechnology laboratories where thousand of assays are performed daily using small volumes of analytes and reagents in small reaction chambers.

Conventional lithography and electron beam lithography are widely used in semiconductor and microelectronics industries to create pattern structures with feature size down to 250nm routinely and with UV-excimer laser (l=157nm) down sizing of features to 100 nm is feasible. However these techniques are intrinsically very expensive, require extremely clean room conditions and require very expensive optics with increasing energy of the UV-light (l=157nm). Moreover, they require very flat surface, involve multiprocessing steps, provide little or no control over the chemistry of the surfaces and hence not very suitable to create surfaces with specific functionalities and not directly applicable to patterning of biological molecules (proteins and cells), which is essential for a number of biological and biochemical analysis. These techniques are, in general, unfamiliar and inaccessible to chemists and biologists and are limited in their applications to biotechnology and biology.

A new set of non-photolithography techniques clubbed under 'Soft Lithography' are fast emerging as alternatives to conventional photolithography for fabrication of micro and nano structures in a variety of metals and polymers.

These techniques also offer great promise to fabricate pattern surfaces with contrasting properties such as wetting, chemical reactivity and surface charges at the micron and submicron level for a variety of applications in various fields.

These high-resolution lithography techniques are based on transfer of a pattern from an elastomeric stamp to a solid substrate by conformal contact. A nanoscale interaction between the substrate and stamp on microscopic scale allows transfer of the material from stamp to substrate to the specified regions that control the surface structure at the molecular level. There are several variants of softlithography. They include: micro contact printing (mCP); micro replica molding (mREM); micro transfer molding (mTM); micromolding in capillaries (MIMIC); solvent assisted micro molding (SAMIM).

These techniques are inexpensive, simple, easy to implement, efficient, single step processing, do not require any sophisticated equipment, do not require any stringent control on the environmental conditions, provide an easy control on the surface topography and functionalities at the molecular level for various biological & biotechnological applications and can be used to pattern planar and curved surfaces.

Key Elements of Softlithography

There are three basic elements of soft lithography:

1. Fabrication of elastomeric stamp with the desired pattern structure on its surface
2. Self-assembled monolayers
3. Preparation of standard masters.

General strategies of soft lithography are shown schematically in Figure 6.1. One of the key elements of this technique is the fabrication of topographically patterned elastomeric stamp. The elastomeric stamp is fabricated by casting appropriate prepolymer against a master containing the desired pattern structure on its surface. The master can be prepared on a glass or a silicon substrate using conventional lithography and etching technique, or by electron beam lithography or by any other standard technique. The prepolymer is subsequently polymerized under suitable conditions and can be peeled off from the master. To facilitate peeling-off the stamp from the substrate, the substrate's surface has to be treated with suitable fluorine-substituted silane coupling agent before casting. The cured elastomeric block contains on its surface the structures that are replica of the pattern structures on the master *i.e.* the raised structures on the master are the depressions in the stamp. The pattern structure on the polymer exhibits high fidelity with dimensional tolerance of a few percent. A number of polymers can be used as 'stamp' for transferring patterns from the masters. The stamp polymer should meet the following requirements: it should be durable and exhibit dimensional stability; should not deform irreversibly under

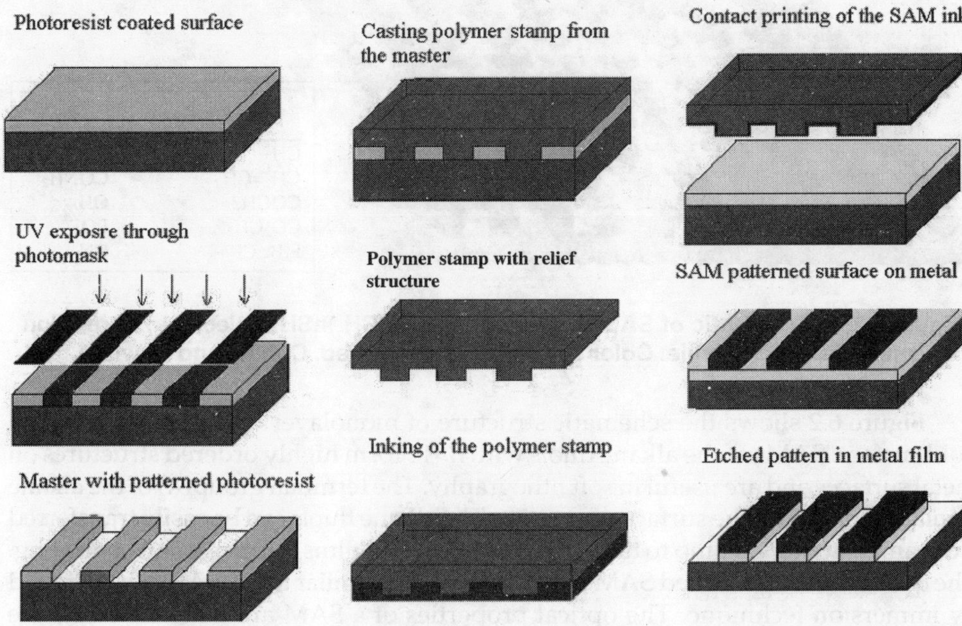

Figure 6.1: Schematic Diagram of the Basic Elements of Soft Lithography (μCP)

pressure; should not swell under different solvents; should have low surface energy (~20 dynes/cm^2); should exhibit high tensile modulus (2MPa). The high tensile modulus can be increased to 8MPa by suitably modifying the polymer ingredients. Poly (dimethyl siloxane) (PDMS) from Dow Corning USA is one of the most widely used polymers for stamp formation. The same master can be used to prepare elastomeric stamps repeatedly and each stamp can be used to print large number of impressions without sacrificing quality.

A number of materials can be used as 'ink' solutions that adsorb irreversibly to the elastomeric stamp and subsequently can be transferred to the specified regions of different substrates. As 'ink' once can use a range of self-assembly monolayer forming solutions, colloids, inorganic solids, proteins and cells. Alkane thiols [X (CH_2)n SH] are most widely used to fabricate pattern structures on various metal films like gold, silver, copper etc. The terminal group X can be fluorocarbons, acids, esters, amines, alcohol, or ethers. Alkane thiols form highly ordered structure films by spontaneously bonding of the sulfur atoms on to the metal surfaces and the alkyl chains are trans extended due to van der Waals interaction and tilted at ~ 30° from normal to the surface.

$$X\,(CH_2)nSH + Au^0 \longrightarrow X\,(CH_2)nS^-Au^1 + \tfrac{1}{2}\,H_2$$

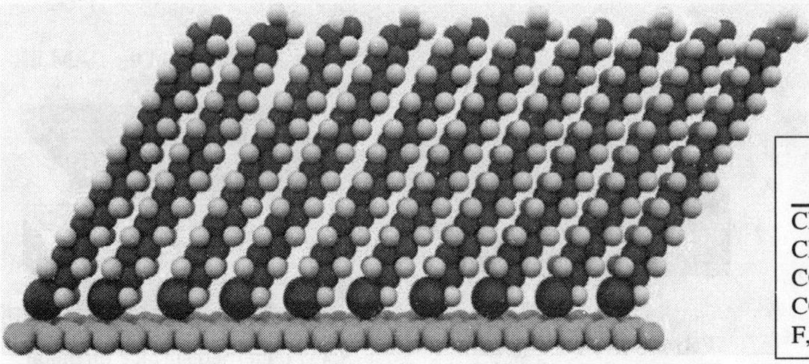

X	
CH_3	EG_n
$CH_2{=}CH_2$	$CONH_2$
COCH3	OH
CO_2CH_2	PO_4^{-2}
F,Br,Cl	CN

Figure 6.2: A Schematic of SAM (n-alkanethiol $CH_3(CH_2)nSH$ molecules) Formation on an Au(111) sample. Color code Au: yellow, S: red, C: blue and H: white.

Figure 6.2 shows the schematic structure of monolayer of alkane thiolates on gold surface. SAMs of the alkane thiols with n>11 form highly ordered structures on metal surfaces and are useful in soft lithography. The terminal group(X) of the alkane thiolates dominates the surface properties. The alkane thiols can be easily transferred from an elastomeric stamp to the surface of thin metal films by microcontact printing. The quality of the patterned SAM monolayer is very similar to the SAM layer formed by immersion technique. The optical properties of a SAM are determined by the thickness of the gold film. The SAMs of alkyl siloxanes can be formed on hydroxylated surfaces of glass, Si/SiO_2, quartz, oxidized polymer surfaces. These SAMs exhibit many interesting features that include ease of fabrication, good stability in ambient laboratory conditions, and in controlling the interfacial properties. Besides, they provide excellent systems to study wetting, adhesion, corrosion, lubrication, selective nucleation and growth, protein adsorption and cell attachment. The monolayer of alkane thiolates act as nano thick resist in lithographic process and some of the concerns like depth of focus, optical transparency in UV and DUV regions, shadowing and undercutting that currently limits the performance of photoresist in high resolution imaging are not important. The patterns in SAMs can be transferred into thin films of metal using appropriate wet etching solutions. For etching patterns in metal films of gold and silver, etching solution uses thiosulfate as coordinating ligand and ferri/ferro cyanide solution as oxidant. The following chemical reactions take place at the metal surface:

$$Au^0+2S_2O_3^{2-} = Au(S_2O_3)_2^{3-} + e^-$$

$$Au^0+2\,OH^- = Au(OH)_2^- + e^-$$

$$Au(OH)_2^- + 2S_2O_3^{2} = Au(S_2O_3)_2^{3-} + e^-$$

$$Fe(CN)_6^{3-} + e^- = Fe(CN)_6^{4-}$$

This etchant generates fewer defects in the SAM covered regions and yields microstructures with high edge definition. Figures 6.3 and 6.4 show optical photograph and SEM pictures of micro patterns formed on gold film formed by micro contact printing of hexadecane thiol (HDT) and subsequent etching in ferro/ferricyanide solution. This pattern contains feature size down to a few micron.

Figure 6.3: Optical Micrograph of a Micro Pattern Formed in a Gold Film

It can be clearly seen that the pattern structures with dimensions down to a few hundred-nano meter could be created in gold/silver film. The edge resolution is controlled by the diffusion of alkane thiols during stamping which in turn is dependent on the molecular weight of the alkane thiol. Octadecanethiol (ODT) is well suited for patterning nano-scale features instead of HDT.

The patterns in metal films can be used as secondary masks for etching underlying layers of silicon dioxide and silicon or can be used as microelectrodes and diffraction grating.

Patterned Surfaces with Contrasting Wetting Properties for Selective Nucleation and Growth

SAMs with different functional groups at the termini (fluorocarbons, acids, esters, amines, alcohol, ethers) of the alky chain of the alkane thiols can be co-patterned to create pattern structure with contrasting wetting, dewetting properties. In the first place, a monoalyer of alkane thiol with methyl terminal group (HDT) is transferred to the selective regions of the gold/silver coated substrate by mCP. Subsequently the underivatized metal-coated regions are coated with a monolayer of mercapto hexadecanoic acid by immersing the whole substrate in a 0.01m ethanolic solution. Such a pattern structure shows highly hydrophobic regions (derivatized with methyl terminal group) and highly hydrophilic regions derivatized with acid group.

Figure 6.4 shows two pattern structures showing different wetting and dewetting regions. The hydrophilic regions are wetted by water while hydrophobic regions form water droplets.

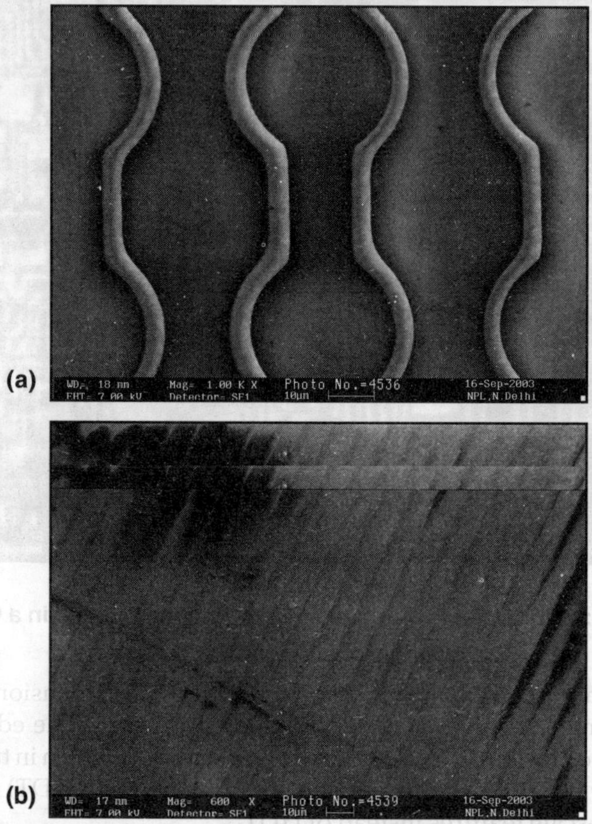

Figure 6.4(a,b): SEM Photographs of Patterns Created in Gold Film

Such patterned structures can be used for selective nucleation and growth of inorganic crystals; optical wave guides and micro lenses of poly urethanes; selective adsorption of proteins and cells; selective chemical vapor deposition on Si/SiO_2 patterned with SAMs of siloxanes. Figure 6.6 shows the selective growth of NaCl crystals on the hydrophilic regions.

Soft Lithography for Biological Applications

Microfabrication is equally important and lucrative for various biochemical and biological analysis. Thousands of assays are performed daily in pharmaceutical and biotechnology laboratories. It is necessary to develop diagnostic tools that use small quantities of analytes and reagents. It requires convergence of microfabrication, automation sciences and biology. Silicon micromachining and MEMS were the first to be adopted for the fabrication of microstructures for biological research.

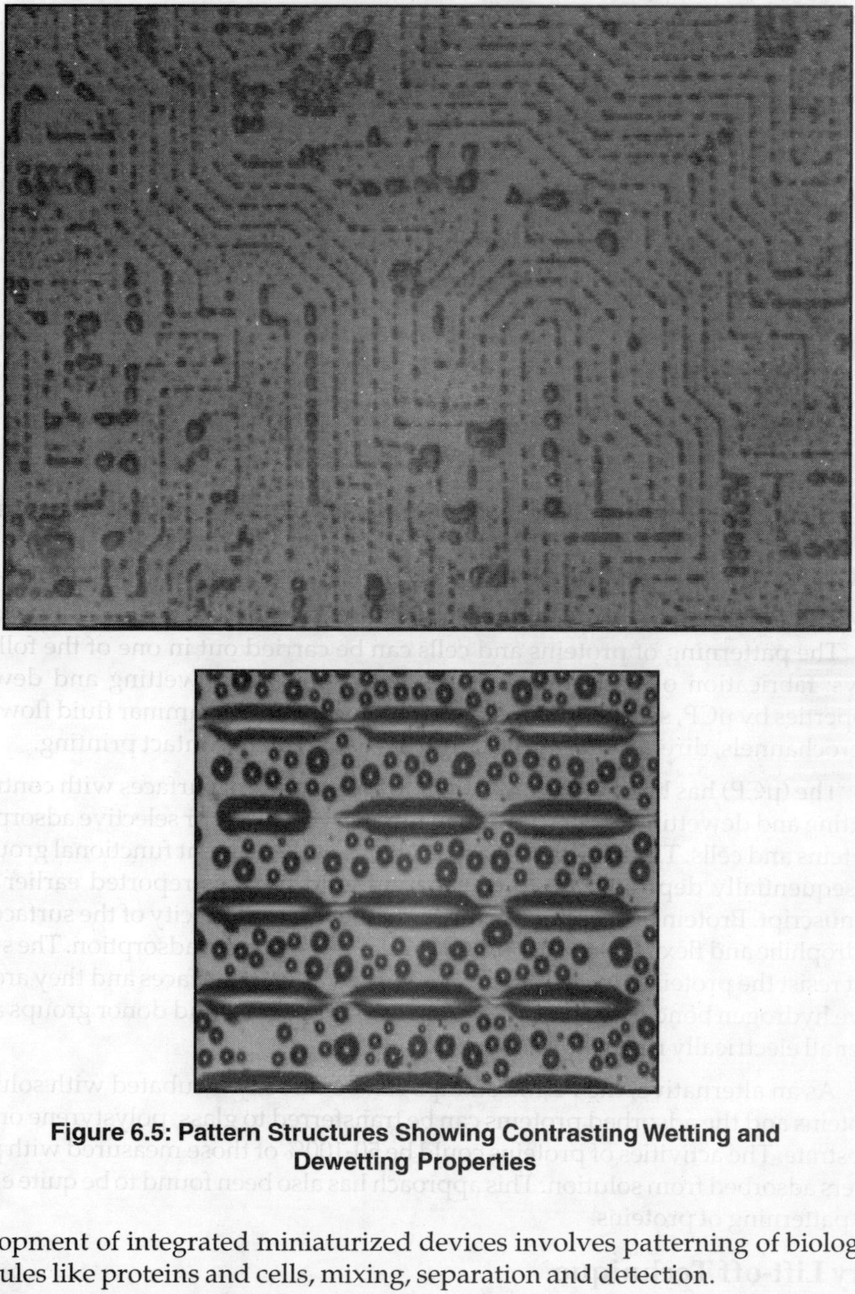

Figure 6.5: Pattern Structures Showing Contrasting Wetting and Dewetting Properties

Development of integrated miniaturized devices involves patterning of biological molecules like proteins and cells, mixing, separation and detection.

The SAMs of alkane thiols are best suited for functionalization and patterning of surfaces (control of surface biochemistry) required by many applications in biology and biochemistry. Selective adsorption of proteins to the specified regions is central to many systems of biotechnology such as biosensors, implants, chromatographic

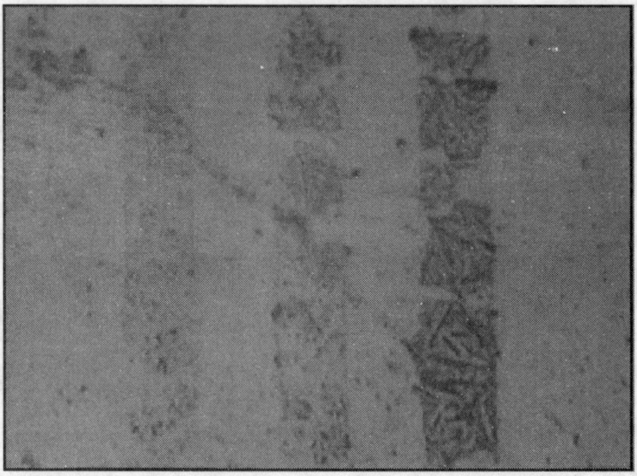

Fgure 6.6: Selective Growth of NaCl Crystals on the Hydrophilic Regions

and electrophoretic media, containers for storing and transferring proteins, tissue engineering and fundamental studies of cell biology.

The patterning of proteins and cells can be carried out in one of the following ways: fabrication of patterned surfaces with contrasting wetting and dewetting properties by μCP, selective deposition of biomolecules by laminar fluid flow inside microchannels, direct patterning of proteins using micro contact printing.

The (μCP) has been very convenient to create pattern surfaces with contrasting wetting and dewetting properties and in turn very suitable for selective adsorption of proteins and cells. The SAMs of alkane thiolates with different functional groups can be sequentially deposited on to transparent gold films as reported earlier in this manuscript. Protein adsorption correlates to the hydrophobicity of the surface while hydrophilic and flexible polymer (PEG) surface resist protein adsorption. The surfaces that resist the protein adsorption are also termed as inert surfaces and they are polar, have hydrogen bond acceptor groups, have no hydrogen bond donor groups and are over all electrically neutral.

As an alternative, the PDMS stamps can be directly incubated with solution of proteins and the adsorbed proteins can be transferred to glass, polystyrene or silicon substrate. The activities of proteins could be 50-100% of those measured with protein layers adsorbed from solution. This approach has also been found to be quite effective for patterning of proteins.

Dry Lift-off Technique

Besides the above listed techniques, selective deposition to the specific region could be achieved by blocking undesirable regions with elastomeric membranes containing through holes. Such membranes can be formed by spin coating PDMS on a master of photoresist posts to thickness less than the heights of the posts. On curing, PDMS membrane can be removed from the master as a freestanding film with features in the shape of the posts on the master. These membranes can be useful in patterning metals and proteins. The membranes can be bonded to the HDT coated gold surface and after adsorption of proteins the membrane can be removed to yield patterned protein surface.

Patterning of Biomolecules Using Laminar Fluid Flow Inside Microchannels

The fluid flow inside confined microfluidic channels provides an equally important and convenient method for selective deposition of biomolecules. Micromolding of soft polymers using soft lithography technology provides a very simple technique to create micro channels with varying dimension ranging from a few to several hundred micron. The polymer block containing channels is sealed reversibly or irreversibly against flat surfaces. The PDMS surface can be oxidized by treating under plasma-oxidation leading to the formation of SiOH groups on the surface. The two oxidized surfaces adhere irreversibly on bringing in contact with

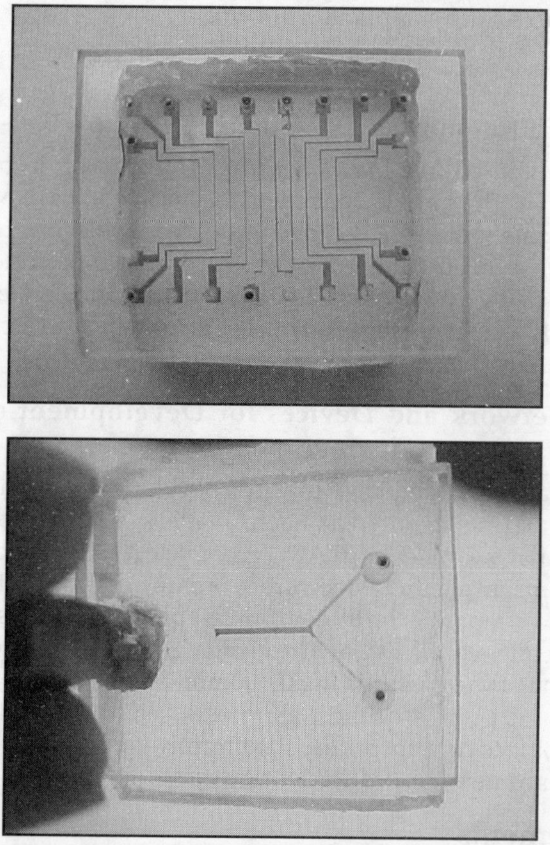

Figure 6.7: Controlled Fluid Flow Inside Microchannels with Independent Access Ports. The channel width varied from 40 to 150 micron.

each other presumably by dehydration of SiOH groups. Fluids containing the species to be patterned flow through the channels and pattern the bound surfaces through covalent attachment, adsorption or the dissolution of the material already present. The multi-channels allow for simultaneous patterning of different proteins in well-defined regions adjacent to each other. In yet another design, the surface can be non-specifically patterned with collagen or fibronectin on a variety of substrates and then cells can be seeded on the patterned surfaces. This technology is going to be extremely useful in development of biosensors and analytical devices in future.

Advantages of Polymer Based Microfluidic Devices

- Replication of micron size features with high fidelity with ease and rapidity
- They are cheap and rugged
- Requires ambient clean room facility for their fabrication
- Sealing is done at ~100°C and can be reversible or irreversible.
- Microfluidic devices thus prepared would be good for multiple uses.
- They are nontoxic and mammalian cells can be cultured directly on to them and devices can be implanted in vivo.

Future Plans

Fractionation of Proteins and DNA using Capillary Gel-Electrophoresis

The channels will be filled with agarose (1% solution in buffer) at constant temperature (~60 °C). The sample containing the nucleic acids (DNA or RNA) will be loaded. The channels will be sealed and subjected to electrophoresis in order to separate the nucleic acids of different sizes. Fractionation of proteins and DNA can also be carried out using acrylamide gels in the microchannels. Detection of proteins can be done using fluorescence tags and nucleic acids will be detected using DNA gel documentation system.

Microfluidic Network and Devices for Development of Miniaturized Integrated Devices

There is a great deal of emphasis towards development of miniaturized devices that perform several tests required in macro analysis. Such devices and systems are called micro total analysis systems (μTAS). A complete μTAS should be able to perform all tasks including sample and concentrate organisms from a fluid, lyse them to release nucleic acids, separate the DNA from rest of the cell debris, amplify the DNA using PCR and sequence the DNA. The challenges are to create systems that can manipulate laminar flow of fluids in 50-100mm channels; functional components like active valves and pumps, connectors, mixers, separators, detection systems etc. Microfluidics is going to contribute very significantly for development of miniaturized integrated devices as next-generation bioassays.

Acknowledgements

I would like to thank my colleagues Dr. P. K. Singh, Dr. Mrs. Daman Saluja (Delhi University), Mr. Vikash Sharma, Mr. Vinod Tanwar for their contributions during the various stages of this work. I would also like to thank Dr. Vikram Kumar, Director NPL for his constant support and encouragement. I would like to acknowledge DST for Grant:DSY/TSG/ME/2002/13.

References

- Bernard, A., Delamarche, E., Schmid, H., Michel, B., Bosshard, H.R., Biebuyck, H.A. (1998). Printing Patterns of Proteins, *Langmuir*, **14**: 2225-2229
- Delamarche, E., Bernard, A., Schmid, H., Bietsch, A., Michel, B. and Biebuyck, H.A (1988). Microfluidic Networks for Chemical Patterning of Substrates: Design and applications to Bioassays, *J. Am. Chem. Soc.* **120**: 500-508
- Duffy, D.C., McDonald, J.C., and Schueller, O.J.A., (1988). Rapid Prototyping of Microfluidic Systems in Poly (dimethylsiloxane), *Anal. Chem.*, **70**: 4974-4984
- Kumar, A., Biebuyck, H.A., and Whitesides, G.M (1994). Patterning self-assembled Monolayers: Applications in Materials Science, *Langmuir*, **10**: 1408-1511
- Kumar, A., Abbott, N.L., Hans, A. K., Biebuyck, H.A., and Whitesides, G.M (1995). Patterned self-assembled Monolayers and Meso-Scale Phenomena *Acc. Chem. Res.*, **28**: 219-226
- McDonald, J.C., and Whitsides G.M, (2002). Poly (Dimethylsiloxane) as a Material for Fabricating Microfluidic Devices, *Accounts of Chemical Research*, **35**: 491-499.
- Mrksich, M., Duke, L.E., Tien, J., Ingber, E., and Whitesides, G.M (1997). Using Microcontact Printing to Pattern attachment of Mammalian Cells to self-assembled Monolayers of Alkane Thiolates on Transparent Films of Gold and Silver, *Experimental Cell Research*, **235**: 305-313
- Pamela, M. and Craighead, H.G. (1996). Microcontact Printing and Pattern Transfer using Trichlorosilanes on Oxide Substrates, *Appl. Phys. Lett.*, **66**: 1022-1024
- Wilbur, J.L., Kumar, A., Biebuyck, H.A., Kim, E., and Whitesides, G.M (1996). Microcontact Printing of self-assembled Monolayers: Applications in Microfabrication, *Nanotechnology*, **7**: 452-457
- Xia, Y., Zhao, X.M., and Whitesides, G.M (1996). Pattern Transfer: self-assembled Monolayers as Ultrathin Resists, *Microelectronic Engineering*, **32**: 255-268

7

Current Mode Approach for Low Voltage Analog Designs in Wireless Applications

S.S. Jamuar

Electrical and Electronic Engineering Department, Faculty of Engineering, Putra University, Malaysia, 43400

E-mail: ssjamuar@eng.upm.edu.my

ABSTRACT

The importance of power consciousness is well understood by many in wireless applications. With the growing experience on the fundamental principles of power conscious design, current mode approach for low voltage analog designs are now becoming known. Current mode approach has proven to be a better alternative for a low voltage high performance analog circuit design. This paper covers the current trends of design techniques for low voltage analog circuit designs utilizing current mode approach within the CMOS environment. Some of the new techniques are decoupling scheme, improved current mirror, transistors operating in sub-threshold region, active cascode scheme and controllable gain. This paper presents the design techniques applied on the building blocks (i.e LNA, mixer, filter, VCO and VGLNA) of a typical wireless receiver system including RF circuits using MEMS technology. These techniques are expected to find use in most analog circuits emphasizing on low power consumption especially in the wireless applications.

Keywords: *Current mode approach, Low voltage, Wireless applications, CMOS technology, MEMS technology.*

Introduction

The era of wireless communications is expanding in an unprecedented way and such developments have prompted many design challenges specifically in low-voltage and low power integrated circuits. Most of the key issues in designing the wireless system are integration of the digital base band system with the high frequency analog front end, and secondly the low voltage or low power circuits. The development of low power microelectronics had since begun with the invention of transistors years ago.

High frequency analog circuits are typically implemented with a GaAs or bipolar devices/technology, while the base band digital signal processor (DSP) is implemented within a complementary metal oxide semiconductor (CMOS) technology. However, over the recent years, there has been a vast interest focusing on the high-frequency analog front-end components, which are designed within the CMOS environment. The suitability and cost effectiveness of CMOS technology for the design and development of digital circuits has helped accelerate the advancement and maturity of this technology. Tremendous advances in CMOS technology have shown no sign of slowing down. The current commercially available minimum size channel length of CMOS process is 0.18 mm, compared to the 2.0 mm state of the art technology that was available in 1983.

There has been significant amount of effort being put in implementing the analog front end in wireless applications within the CMOS environment. One of the main issues in the wireless applications is low voltage and low power at high frequency operation. One common practice to reduce the power consumption of a circuit is to scale down the supply voltage but this would results in degrading the performance of the circuits in terms of available bandwidth and voltage swing. Alternatively, decreasing the threshold voltage of MOSFETs would reduce the performance loss but results in increased static power dissipation. The analog circuits benefit marginally from the scaling, as the minimum size transistors cannot be used in analog circuit structures due to the noise and offset voltage constraints. The designers are looking for ways to design efficient low power analog circuits. The low voltage and low power operation complicates the design of the circuits. Simple circuits having smaller number of MOSFETs are expected to give better circuit performance because of small device and stray capacitances but in reality, it works in opposite way. Therefore, it is highly desirable to design simple and efficient circuit structures for low voltage analog circuits.

Portability of electronic systems is also an issue in view of the need for low power systems which include battery-operated products like cell-phones, hand held multimedia terminals, hearing aids and implantable cardiac pacemakers, etc. The increasing demand of portable equipment to have reasonable battery life and weight has encouraged research and development efforts in design and operation of low voltage circuits with low power consumption. In addition, size factor would be another issue to be conferred in a portable system.

Conventional design techniques for low voltage in analog circuits differ considerably from the high voltage analog circuit design techniques. All conventional analog circuits are voltage mode circuits (VMCs) and these circuits suffer from the drawback that the output voltage does not change instantly with changes in the input voltage and does not have high voltage swings. For better signal to noise ratio (SNR), higher supply voltages are required. Thus, VMCs are not suitable for high frequency applications.

Current mode circuits (CMCs) where the input/output signals are primarily currents and the currents determine the complete circuit response, are suitable for integration in CMOS technology. CMCs have simple architecture and their operations are independent of the supply voltages. With the proper biasing conditions of current mode circuit elements established, the circuits would have improved performance. The current mode approach provides a better alternative for low voltage high performance analog circuit design where the only concern for the circuits is the current levels.

The recent developments in wireless communication have motivated much interest as well as design challenges in low voltage and low power integrated circuits. Each building blocks of a portable wireless device contribute significantly to the overall power consumption of a receiver system. Thus, stringent requirements for different wireless systems place particular emphasis on each building blocks of a receiver (Figure 7.1). There are endless efforts in developing different design techniques for different building blocks in the wireless system suitable for low power operation.

This article is aimed at providing a brief overview of the current mode approaches in developing low voltage design techniques prevalent today for analog circuits in the wireless applications. It would also cover some of the revolution in the microelectromechanical system (MEMS) technology for wireless applications. Due to the system's flexibility, the technology can be exploited to overcome the limitations exhibited by integrated RF devices. In what follows are different analog design techniques being implemented and the applications of the design techniques on building blocks of a wireless system. Different design techniques are discussed and implemented respectively with emphasis on low power dissipation.

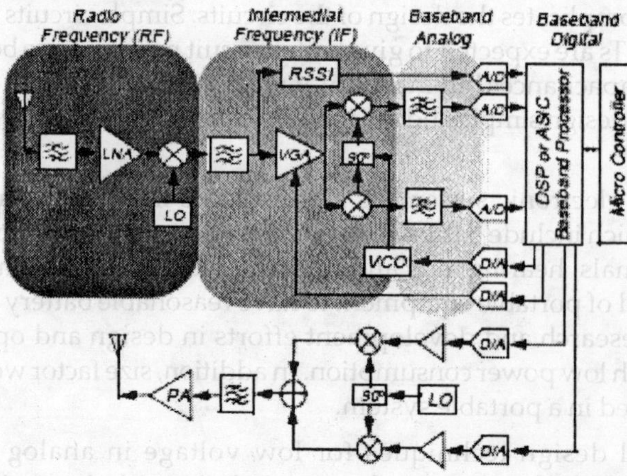

Figure 7.1: Block Diagram of a Generic Wireless Transceiver

Approaches for Low Voltage Analog Designs

This section provides some insights using current mode approach in designing low voltage RF front-end circuits. Current mode approach proves to be a better alternative where the designer would be more concerned of the current levels in the operation of the circuits while keeping the supply voltage low and constant.

This section explores the design techniques for wireless applications such as decoupling scheme, improved multiplier based on Sackinger current mirror, MOSFETs being biased to operate in sub-threshold region, active-cascode transconductor scheme and a controllable gain for a variable gain low noise amplifier. Capability of microelectromechanical system (MEMS)-based circuits used in large quantities could enhances the robustness of the wireless system and substantially reduce power consumption as well.

Decoupling Scheme

In radio frequency (RF) integrated circuit (IC) functional blocks, many circuit topologies use dc current sharing to realize the necessary biasing, functionality and/or performance. One example of such an RF IC functional block is a low noise amplifier (LNA) (Figure 7.2), which can be realized in CMOS technology environment as well.

The cascode RF amplifier structure increases stability and allows matching for noise as well as input impedance. This amplifier requires two stacked transistors for functionality and if each transistor has a turn-on voltage of 0.8 V, the absolute minimum supply voltage is 1.6 V. For the classic cascode amplifier, both dc and ac currents are shared between the two transistors. If the ac and dc portions of the circuit could be decoupled, the power supply voltage could be reduced.

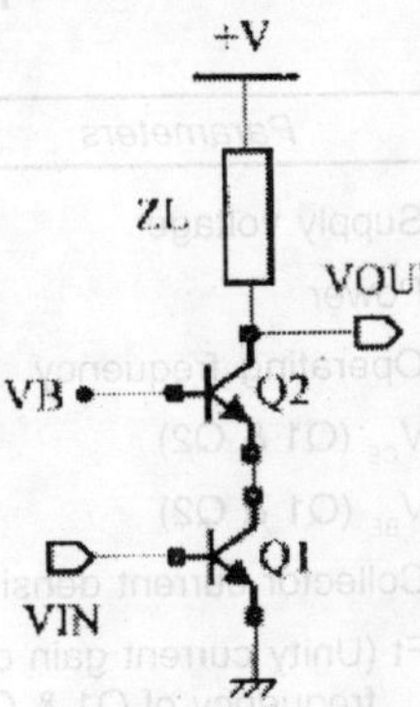

Figure 7.2: Classical Cascode Amplifier Using Bipolar Devices

A proposed decoupling scheme, which uses two on-chip *LC* tanks and one coupling capacitor, were used. As shown in Figure 7.3, the *LC* tanks or "RF traps" in the proposed scheme, provides low impedance across its terminals at dc and relatively high impedance at RF. The coupling capacitor is used to couple the RF energy between the two elements. In this proposed scheme, no dc current is shared between the two transistors. So, the minimum voltage supplied required is only 0.8 V, not 1.6 V if the transistor has a turn-on voltage of 0.8 V.

The decoupling topology has been applied on a low noise amplifier where the transistors were all biased to the operating frequency of 1.9 GHz (Figure 7.4). A summary of comparison was made between proposed schemes with a 2 V conventional cascode amplifier structure for LNA as given in Table 7.1.

As shown, there is an enhancement of noise figure (NF), 0.22 dB attributed by the fact that some of the shot noise associated with Q2 is diverted to ground by the

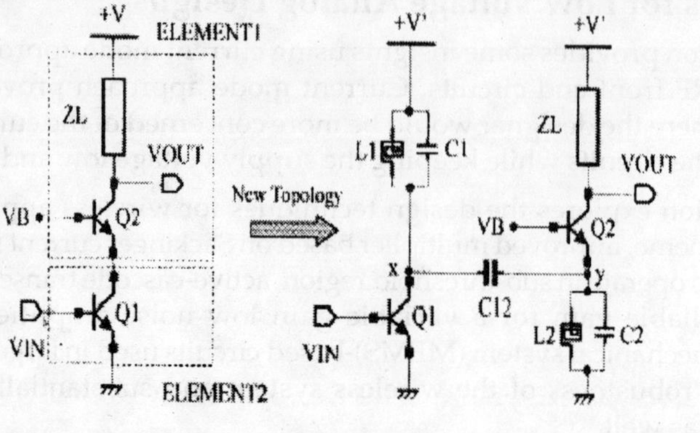

Figure 7.3: New Cascode Amplifier Using Capacitively Coupled *LC* Tanks

capacitively coupled resonating elements and does not enter *Q1*. Another advantage of this low voltage cascode structure is the collector emitter current of each transistor can be independently controlled. Thus, this could help increase linearity and reduce the noise figure.

Table 7.1: Comparison Between a Traditional Cascode Amplifier and a Low Voltage LNA

Parameters	*Traditional Cascode*	*Low Voltage Cascode*
Supply voltage	2 V	1 V
Power	2.8 mW	2.8 mW
Operating frequency	1.9 GHz	1.9 GHz
V_{CE} (Q1 & Q2)	1 V	1 V
V_{BE} (Q1 & Q2)	0.81 V	0.81 V
Collector current density Q1 & Q2	16 μA/μm²	16 μA/μm²
Ft (Unity current gain cut of frequency of Q1 & Q2)	8 GHz	8 GHz
NF50	2.19 dB	1.97 dB
NFmin	2.18 dB	1.96 dB
S11	-29.2 dB	-26.0 dB
GA	9.6 dB	9.6 dB
IP3	- 6 dBm	-6 dBm

The same decoupling topology can also be applied on active mixer circuits. Figure 7.5 shows the proposed topology being applied on a Gilbert mixer along with its standard counterpart. For the mixer structure, the same basic conditions need to be satisfied and the *LC* tanks need to be resonated at RF frequency.

Improved Multiplier Based on Sackinger Current Mirror

Analog multiplier is a useful building block in the realization of functions such as automatic gain control, modulation, filtering, etc. Various CMOS current multiplier

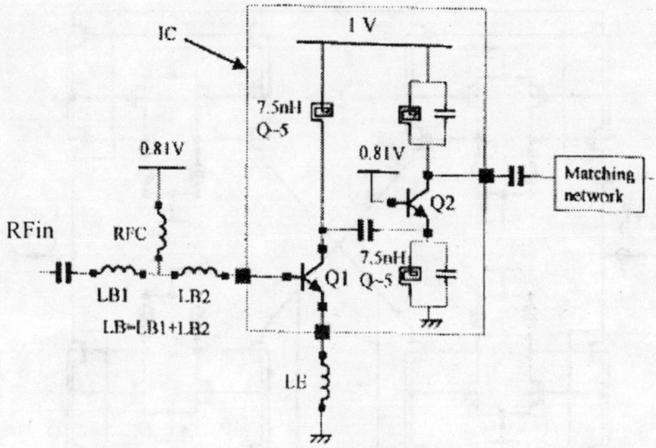

Figure 7.4: Design of Low Noise Amplifier (LNA) Using Decoupling Scheme

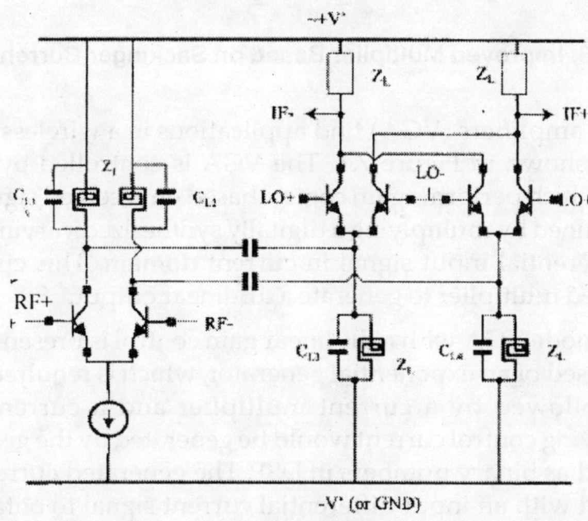

Figure 7.5: Low Voltage Version of a Gilbert Mixer

circuits capable of realizing the product of two currents exist in the literature. Here, a multiplier circuit based on Sackinger current mirror is being considered.

Figure 7.6 shows the improved multiplier, which is made based on a back-to-back connection of two current mirrors implemented with a Sackinger current mirror. Compared to a regular cascode current mirror, Sackinger current mirror allows a greater voltage swing at its input node, an important feature in a low voltage design. This very compact and power efficient CMOS circuit offers a superb exponential characteristic, which is ideally suitable to be used in a variable gain amplifier (VGA) circuit.

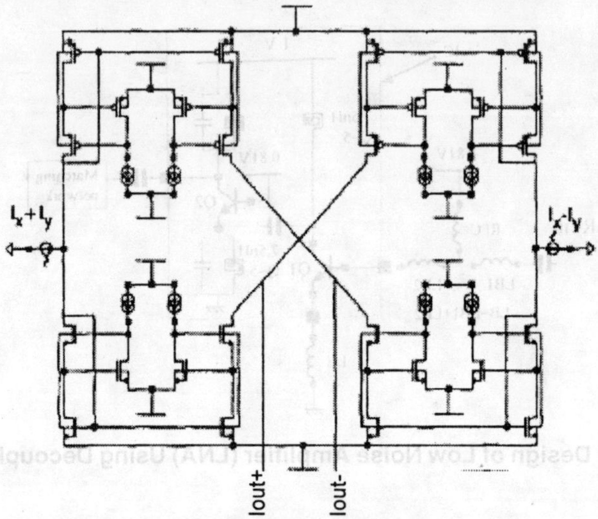

Figure 7.6: Improved Multiplier Based on Sackinger Current Mirror

Variable gain amplifiers (VGA) find applications in a wireless communication receiver chain as shown in Figure 7.1. The VGA is controlled by a digital signal processor (DSP), which performs gain control based on received signal strength. The gain control is obtained by multiplying a digitally synthesized varying control current signal with a differential input signal in current domain. This circuit utilizes the proposed improved multiplier to generate a dB-linear output.

The current-mode VGA with a dB-linear gain control is presented in Figure 7.7. The circuit composed of an exponential generator, which is required to generate the control signal, followed by a current multiplier and a current amplifier. An exponentially varying control current would be generated by the generator based on points represented as binary numbers in DSP. The generated current signal would then be multiplied with an input differential current signal to obtain a differential

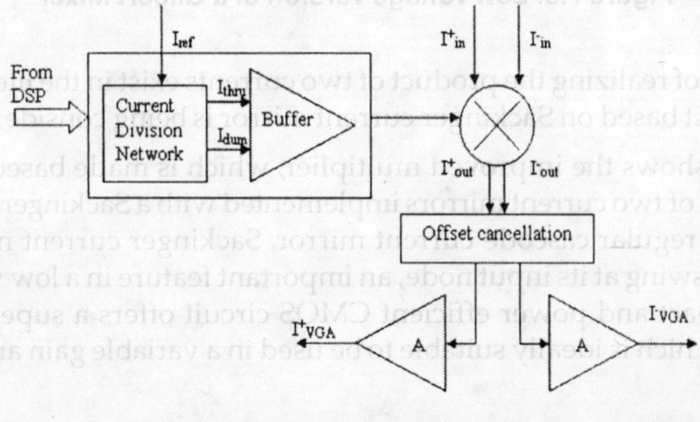

Figure 7.7: Block Diagram of the Variable Gain Amplifier

output current. The differential input signal is scaled to a determined amplitude differential output signal by the exponentially control signal hence realizing a VGA. The DSP changes the digital control word by varying the control current based on any changes in the output signal level.

For the prototype of the VGA, a low supply voltage of 0.9 V is proved to be possible with the improve multiplier based on Sackinger current mirrors. The static power consumption of the VGA was less than 5 mW.

Voltage Controlled Oscillators (VCO) using MOSFETs Biased in Sub-threshold Region

One good application for this kind of MOSFETs in the wireless technology is on a voltage-controlled oscillator (VCO). It is used for modulating the base band signals in portable wireless telemetry devices. The VCO should be able to operate under unusual supply voltage and power requirements as this is important whereby the battery drainage would cause serious problem to a patient.

The VCO circuit presented in Figure 7.8 is based on a modified negative transconductance ($-g_m$) *LC* oscillator topology. The circuit has better phase noise performance and a *LC* tank consisted of L_s and C_s, which is tuned to minimize the effect of the common-mode second harmonic at the output node. The cross-coupled pair of MOSFETs is biased in sub-threshold region. The widths of the transistors are adjusted to give required current which would also provide a transconductance g_m that satisfies the condition of the oscillation given as

$$g_m.R_L \geq 1 \,. \quad (1)$$

The quality factor, Q of the tank circuit is required to be as high as possible in order to get a high equivalent parallel resistance, R_L, which means less g_m is required and thus a lower current is consumed. A high Q enhances the phase noise performance

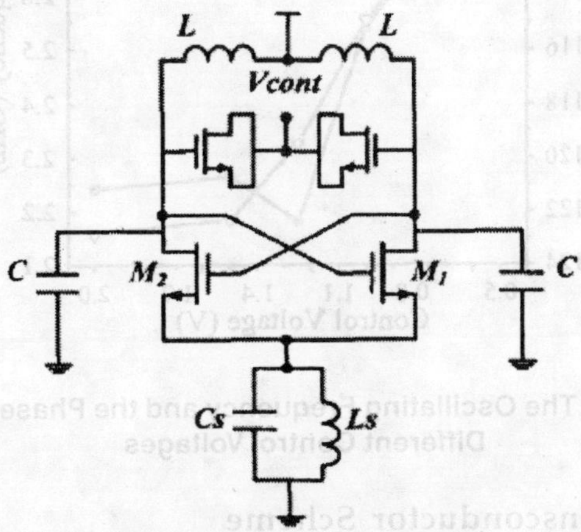

Figure 7.8: VCO Topology Used for Low-voltage and Low Phase Noise

and also, with the biasing current reduced due to high R_L injects less noise to the tank and this leads to better noise performance.

A 2.4 GHz VCO with supply voltage of 0.4 V was simulated using standard CMOS 0.18 μm process. The simulated phase noise output is shown in Figure 7.9 and the phase noise at offsets of 1 MHz, 2 MHz and 3 MHz are -123 dBc/Hz, -129.5 dBc/Hz and -133.1 dBc/Hz respectively. Figure 7.10 shows the frequency and the phase noise plotted for different control voltages. The phase noise shows a minimum value at V_{cont} of 1.3 V because the circuit was optimized at the center frequency corresponding to a V_{cont} of 1.3 V. The simulated power consumption is only 41 μW.

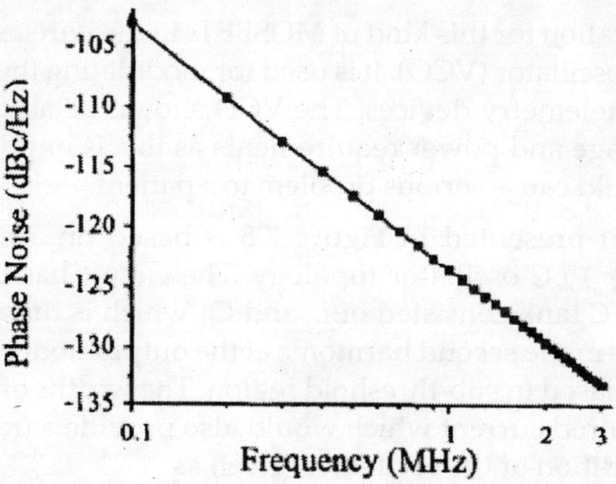

Figure 7.9: Phase Noise at Different Offset Frequencies

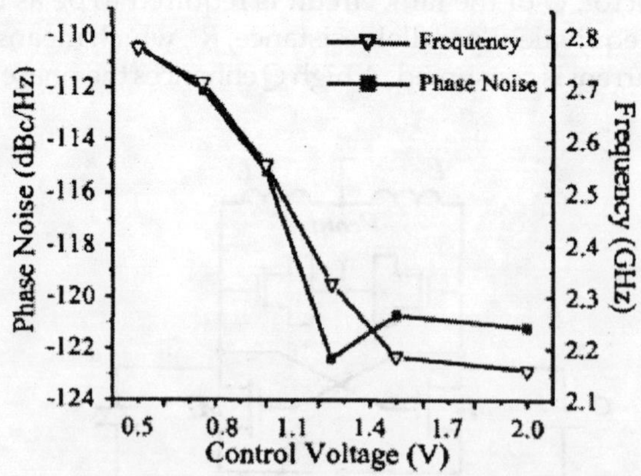

Figure 7.10: The Oscillating Frequency and the Phase Noise for Different Control Voltages

Active-cascode Transconductor Scheme

The proposed active-cascode transconductor as shown in Figure 7.11 composed of a triode-biased transistor *M1* having its drain voltage set according to the control

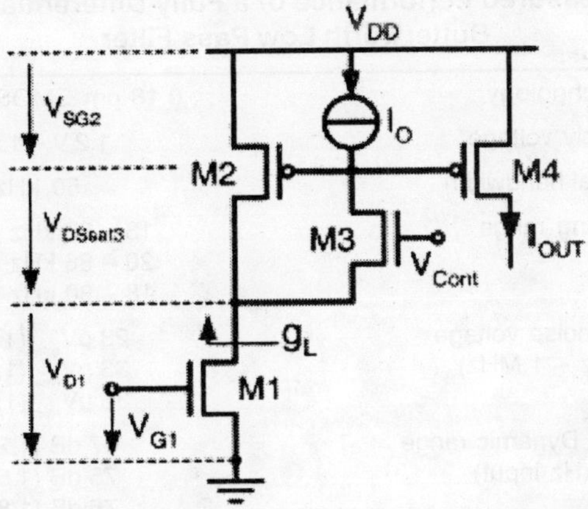

Figure 7.11: Proposed Active-cascode Transconductor Core

voltage V_{cont} and stabilized through the negative feedback loop established by transistors *M2* and *M3*. Transistor *M3* is used to keep the drain voltage of *M1* constant due to its constant current I_O and bias voltage V_{Cont}. The variation in the drain voltage is sensed at the source of common gate amplifier *M3* and amplified at the drain node by its voltage gain. The output voltage of the amplifier *M3* at its drain appears as the input voltage for the common source transconductor *M2*. The drain voltage of *M1* is regulated to V_{cont}-V_{GS3} and the output current of the transconductor is mirrored through transistor *M4*. On the contrary to the regulated-cascode transconductor, the positive control voltage applied at the gate *M3* results in the condition of $V_{cont} < \left(V_{G1} - V_{TO,n}\right)/n + V_{GS3}$ for *M1* to operate in triode region. Thus, the minimum supply voltage is defined as:

$$V_{DD,\min} = V_{D1} + V_{DSsat3} + V_{SG2}, \tag{2}$$

which is generally lower than a regulated cascode transconductor.

The transconductor can be duplicated to realize a fully differential transconductor as shown in Figure 7.12 and using this block a fully differential third-order Butterworth low pass filter can be realized as shown in Figure 7.13. The supply voltage is adjusted from 1.2 to 1.8 V and the measured performance of the filter is summarized in Table 7.2. The filter operates on a supply voltage down to 1.2 V and the 1 %-THD dynamic range of 75 dB can be obtained with 20-kHz input signal at 50-kHz cutoff frequency operating at 1.5 V supply.

The active G_m-C filters based on triode-biased MOS transistors implemented as shown are the choice of baseband filters in the wireless receiver architecture due to its high linearity. The filter exploits the active-cascode scheme with the current of the triode transistor sensed through another path, allowing for regulation of the drain voltage by negative feedback. This topology is simple and appropriate for low voltage and low power operation.

Table 7.2: Measured Performance of a Fully Differential Third-order Butterworth Low Pass Filter

Technology	0.18 μm CMOS process
Supply voltage	1.2 V – 1.8 V
Nominal bandwidth	50 kHz
Tuning range	13 – 65 kHz (1.2 V) 20 – 86 kHz (1.5 V) 18 – 80 kHz (1.8 V)
Output noise voltage (1 kHz – 1 MHz)	28 μV_{rms} (1.2 V) 33 μV_{rms} (1.5 V) 33 μV_{rms} (1.8 V)
1 % - THD Dynamic range (20 kHz input)	67 dB (1.2 V) 75 dB (1.5 V) 76 dB (1.8 V)
1 dB-Compression point dynamic range (20 kHz input)	75 dB (1.2 V) 78 dB (1.5 V) 80 dB (1.8 V)
CMRR (20 kHz)	47 dB (1.5 V)
PSRR+ (DC)	-30 dB (1.5 V)
Power consumption (50 kHz cutoff)	167 μW (1.2 V) 240 μW (1.5 V) 292 μW (1.8 V)

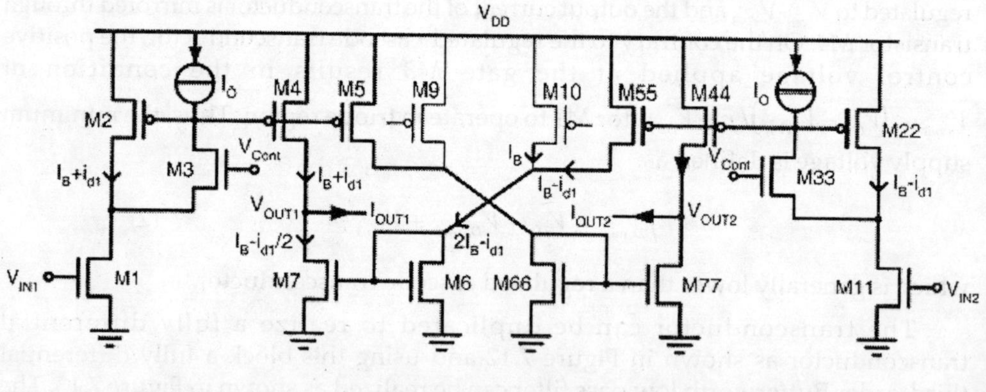

Figure 7.12: Fully Differential Active-cascode Transconductor

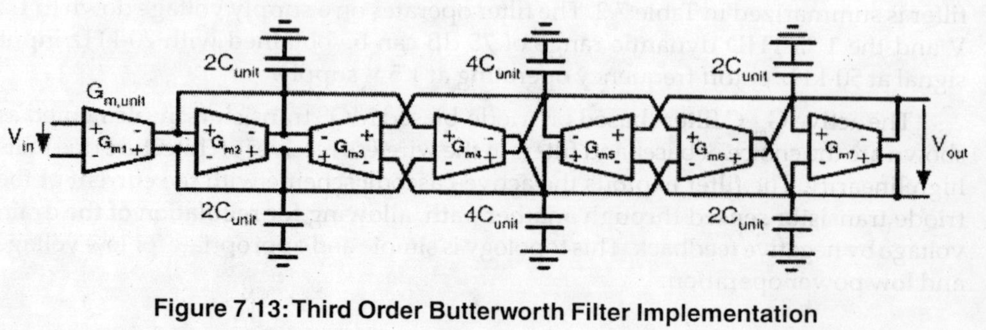

Figure 7.13: Third Order Butterworth Filter Implementation

Variable Gain Low Noise Amplifier (VGLNA)

A simple variable gain low noise amplifier (VGLNA) structure can be implemented using current mirror, at a very low power consumption level. The addition of a controllable gain amplifier is used to prevent saturation of the receiver when the input signal becomes too large. Figure 7.14 is the schematic diagram of a 1 V VGLNA for 2.4 GHz Bluetooth applications.

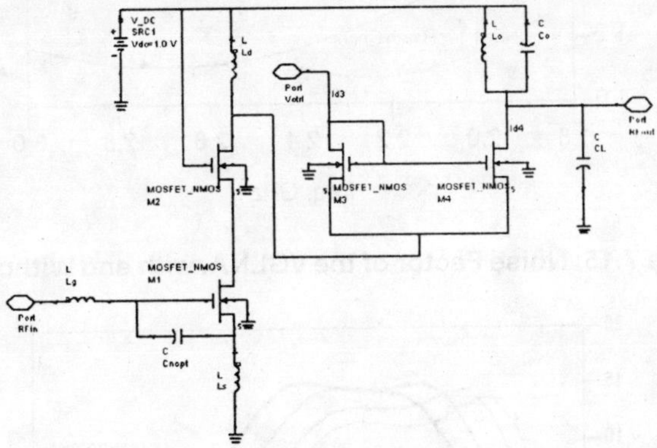

Figure 7.14: Schematic of VGLNA Using Current-mirror Principle

The VGLNA consists of a low noise stage (consists of transistors *M1* and *M2)* and a gain controlled stage (transistors *M3* and *M4*). Both transistors *M1* and *M2* are used in common-source configurations. Transistors *M1* and *M2* share the same bias current, thus, the total power consumption of the current reused amplifier can be minimized. In the second stage of the VGLNA, transistor *M3* would converts the input signal applied as bias voltage, V_{ctrl} to a current signal, I_{d3} by the mediation of impedance at the transistor. The gain of the amplifier is controlled through the geometry ratio of transistors *M4/M3* which induces currents, I_{d3} and I_{d4}. The gain can be varied through the widths, W_3 and W_4. Thus, higher gain is possible at lower current and thus at lower power dissipation.

The VGLNA has been simulated at 2.4 GHz using the standard 0.18 mm CMOS process from TSMC. The simulation results for the VGLNA, which exhibits 1.11 dB noise factor and 15.21 dB gain, are as shown in Figures 7.15 and 7.16 respectively. The VGLNA dissipated 3.09mW of power from a 1 V supply; considering the tradeoffs involved between gain, noise factor and power consumption.

In a typical wireless receiver system, LNA is one of the key components. It tends to dominate the sensitivity of the whole receiver system. With the noise factor and gain being taken care, the implementation of a current mirror in the VGLNA would be simple and appropriate for low power application.

RF MEMS–Wireless Vision

Microelectromechanical systems (MEMS) technology is on the edge of revolutionizing RF and microwave applications. The requirements of the systems are lower weight, volume, power consumption and cost with increased functionality,

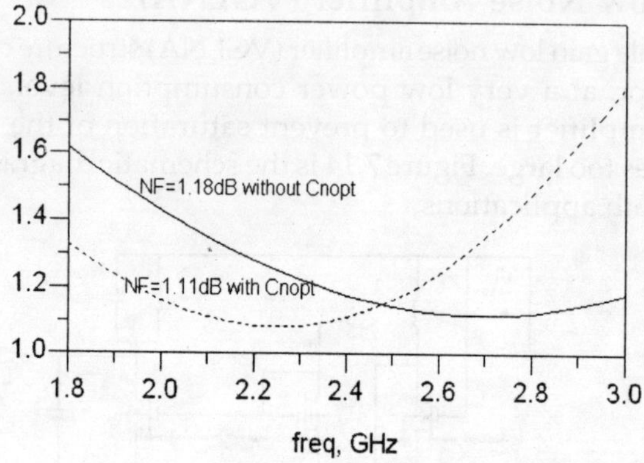

Figure 7.15: Noise Factor of the VGLNA, with and Without C_{nopt}

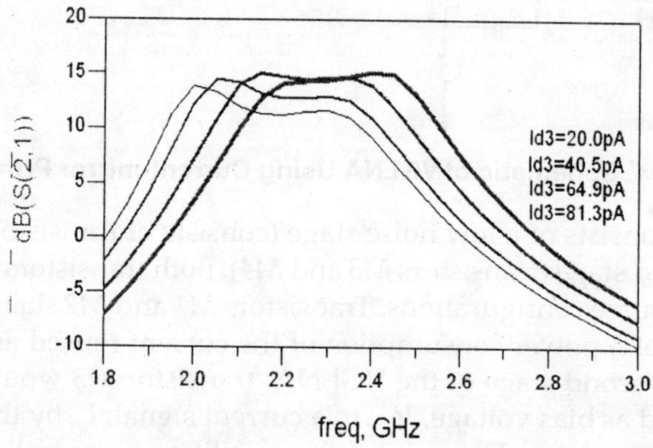

Figure 7.16: Gain, S21 of the VGLNA vrs Frequency for Different Values of I_{d3}

frequency of operation and component integration. The interest in MEMS technology for wireless applications stems from the view that, due to its flexibility, it can be exploited to overcome the limitations exhibited by integrated RF devices. The ultimate goal in applying RF MEMS is to propagate the device-level benefits all the way up to the system level to attain positive levels of system performance, as shown in Figure 7.17.

Some of the circuits exploiting MEMS technology are inductor-based circuits where the inductor played key elements in determining the performance of the tuned circuits such as, impedance matching networks, low noise amplifiers (LNA) and voltage-controlled oscillators (VCO). Figure 7.18 shows an example of a bulk-micromachined inductor in which the substrate has been eliminated from underneath the spiral trace and Figure 7.19 shows a photograph of a tunable RF MEMS inductor emphasized on optimizing the performance of RF front-ends circuits.

Varactors are essential components where the circuit tunability is required, like in variable matching circuits and VCOs (Figure 7.20). Other than inductor-based and

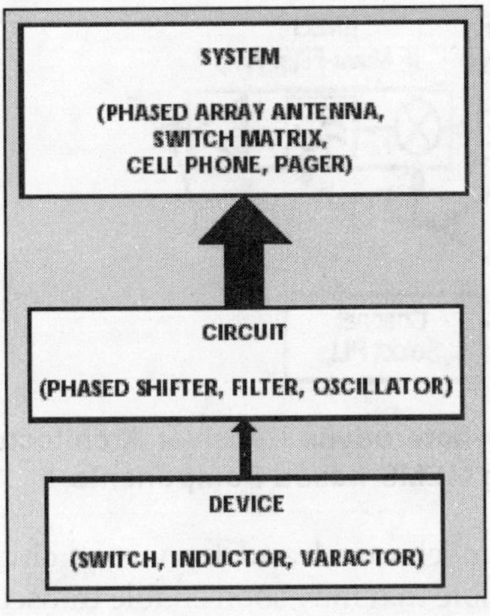

Figure 7.17: Typical RF MEMS Chain from Device to System Level

Figure 7.18: A Bulk-micromachined Inductor

Figure 7.19: A Tunable RF MEMS Inductor

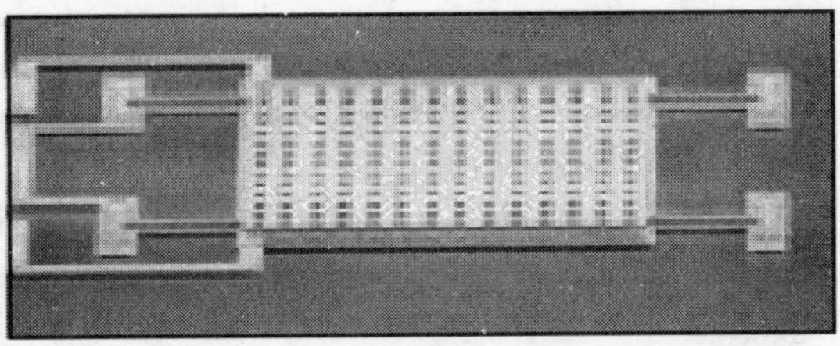

Figure 7.20: A Parallel-plate Varactor

varactor-based circuits, resonators and switches are two components which could benefit from the MEMS technology as well.

With the increasing trend towards more powerful, demanding portable wireless systems, the need for integration is now more acute than ever. The most direct way to harness RF MEMS devices is via direct replacement of off-chip components, as shown in Figure 7.21. A dramatic improvement in receiver noise figure is recorded with direct replacement of the MEMS-based components in the receiver architecture; from 8.8 dB to 2.8 dB. Other advantages of the micromechanical circuits are the micro-scale size and zero dc power consumption. Usage of the components in massive quantities enhances the robustness and trades Q for power consumption.

Micromechanical circuits attained via MEMS technologies have been described to play a key role in scaling the size of communication transceivers. In addition, by

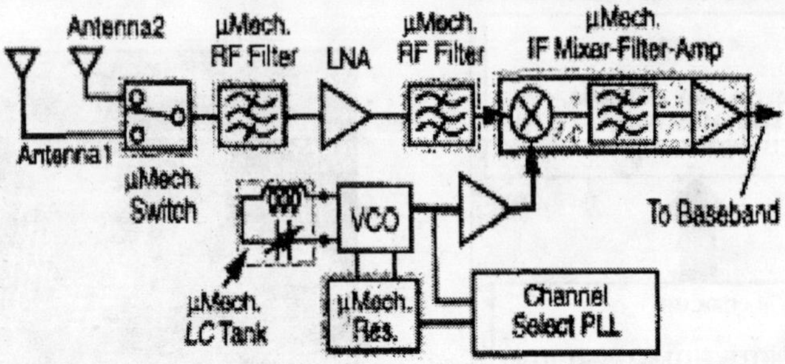

Figure 7.21: System Block Diagram of a Super-heterodyne Receiver Architecture Showing Potential Replacements Using MEMS-based Components

combining the strengths of integrated micromechanical and transistor circuits, previously unachievable functions become possible that may soon enable transceiver architectures especially from the power perspective.

Conclusion

Current mode approach has triggered many promising design techniques in the current trends of low voltage analog design for wireless applications. The approach has shown to be popular and its potentials are outlined in this article. Design techniques like decoupling scheme, improved current mirror, operation in sub-threshold region, active cascode scheme and controllable gain implementation have showcased the benefits of using this approach emphasizing on low power dissipation. In recent year, MEMS technology has also exhibited its flexibility in areas where there is limitation in the integrated RF devices such as micro-scale size and subsequently, low power consumption. Depending upon the nature of the specifications of the receiver system, a necessary design technique can be implemented. If these circuits can be designed to operate at low voltages, then the system in which they are used can also be expected to operate at low voltages, thus dissipating low power.

References

- Abidine, I. Z.-E.-, Okoniewski, M. and McRory, J.G., 2003. A New Class of Tunable RF MEMS Indcutors, *Proceedings of the Int. Conf. on MEMS, NANO and Smart Systems (ICMENS '03)*, 1-2.
- Chandrakasan, P. and Brodersen, R.W., 1995. Minimizing power consumption in digital CMOS circuits, *Proc. IEEE*, vol. 83, No. 4: 498-523.
- Daneshrad, B., 1998. Integrated Circuit Technology for Wireless Communications: An Overview, *IEEE Int. Symp. on Personal, Indoor and Mobile Radio Communications 1998*, vol. 1, 365-369.
- Fakhr, A., Deen, M.J. and deBruin, H., 2004. Low-Voltage, Low-Power and Low Phase Noise 2.4 GHz VCO for Medical Wireless Telemetry, in *2004 Canadian Conf. on Electrical and Computer Engineering*, vol. 3, 1321-1324.

- Geiger, R.L., Allen, P.E. and Strader, N.R., 1990. *VLSI Design Techniques for Analog and Digital Circuits*. Singapore: McGraw-Hill Inc., Ch. 3.
- Gonzalez, R., Gordon, B.M. and Horowitz, M.A., 1997. Supply and threshold voltage scaling for low power CMOS, *IEEE J. Solid-State Circuits*, vol. 32, No. **8:**1210-1216.
- Jamuar, S.S., Sidek, R.S., Khatun, S. and Lee, L., 2005. A 1 V Variable Gain Low Noise Amplifier (VGLNA) for 2.4 GHz application, *IEEE National Symposium on Microelectronics 2005*.
- Kuroda, T. and Sakurai, T., 1995. Overview of low power ULSI circuit techniques", *IEICE Trans. Electron.*, vol. E-78-C, No. **4:** 334-344.
- Langen, K. and Huijsing, J.H., 1998. Compact low-voltage power efficient operational amplifier cells for VLSI," *IEEE J. Solid-State Circuits*, vol. 33, No. **10:** 1483-1496.
- Leeson, D.B., 1966. A simple model of feedback oscillator noise spectrum, *Proc. IEEE*, vol. 54, Feb 1966.
- Liao, C.-H. and Chuang, H.-R., 2003. A 5.7-GHz 0.18-mm CMOS Gain-Controlled Differential LNA with Current Reuse for WLAN Receiver, *IEEE Microwave and Wireless Components Lett.*, vol. 13, 526-528.
- Manku, T., Beck, G. and Shin, E.J., 1998. A Low-Voltage Design Technique for RF Integrated Circuits, *IEEE Trans. on Circuits and System*, vol. 45, 1408-1413.
- Masaki, Akira, 1991. Possibilities of CMOS Mainframe and Its Impact on Technology R & D, *1991 Symp. on VLSI Technology, Digest of Technical Papers*, 1-6.
- Meindel, J.D., 1995. Low power microelectronics: Retrospect and Prospect, *Proc.IEEE*, vol. 83, No. **4:** 619-635.
- MEMS for RF/Microwave Wireless Applications: The Next Wave from http://www.coventor.com/media/papers/.
- Motamed, A., Hwang, C. and Ismail, M., 1997. CMOS exponential current-to-voltage converter, *IEE Electronics Letters*, vol. 33, 998-1000.
- Nguyen, C.T.-C., 2002. RF MEMS for Wireless Applications, *Device Research Conf, Digest*, 9-12.
- Rajput, S.S. and Jamuar, S.S., 2002. Low Voltage Analog Circuit Design Techniques, *IEEE Circuits and Systems Magazine*, vol. 2, 24-42.
- Ravindran, A., Vidal, E. and Ismail, M., 2001. Compact low voltage four quadrant CMOS current multiplier, *Electronics Lett.*, vol. 37, 1428-1429.
- Ravindran, A., Vidal, E., Yoo, S-.J., Ramarao, K. and Ismail, M., 2004. A Differential CMOS Current-Mode Variable Gain Amplifier with Digital dB-Linear Gain Control, *Analog Integrated Circuits and Signal Processing*, vol. 38, 161-174.
- Sackinger, E. and Guggenbuhl, W., 1990. A High-Swing, High-Impedance MOS Cascode Circuit, *IEEE Journal of Solid State Circuits*, vol. 25, 289-298.

- Santos, H. J. De Los, 1999. *Introduction to Microelectromechanical (MEM) Microwave Systems,* Artech House, Norwood, MA.
- Toumazau, C., Lidgey F. J. and Haigh, D. G., 1990. *Analogue IC design: The current mode approach*, Stevenge, U.K.: Peregrines Ltd.
- Yodprasit, U. and Christian, C.E., 2003. A 1.5-V 75-dB Dynamic Range Third-order Gm-C Filter Integrated in a 0.18 mm Standard Digital CMOS Process, *IEEE Journal of Solid-State Circuits*, vol. 38, 1189-1197.

8

RF System for an Ion Source

Luis Miguel Ledo
Specialist,
Centre of Applied Technology and Nuclear Development (CEADEN),
Department of Physics,
P.O. Box 6122, Havana City, Cuba
E-mail: ledo@ceaden.edu.cu

ABSTRACT

An rf application has been introduced to reach controlled plasma production. Facilities of a neutron generator, installed at the Laboratory, are used to support the whole system. An alternative way of exploitation of such a generator is introduced so as to bring it into work on a basis different from the original.

Keywords: Radio frequency, Plasma, Ion source, Impedance matching, RF power.

Introduction

In the Neutron Physics Laboratory at CEADEN, Cuba, there is a neutron generator. Neutron generators are small accelerators consisting of vacuum, magnetic, electrical and mechanical components; radiation sources, cooling circuits and pneumatic transfer systems.

The low voltage (a few 100 kV) neutron generators (IAEA-TECDOC-913, 1996) produce neutrons by the following reaction:

$$^{2}H(d,n)^{3}He \qquad Q = 3.268\ MeV \qquad (1)$$

$$^{3}H(d,n)^{4}He \qquad Q = 17.588\ MeV \qquad (2)$$

The large cross section of the $^{3}H(d,n)^{4}$ He reaction permits high yields of fast neutrons to be obtained even at low deuteron energy.

The process of neutron production on this machine begins with the ionization of deuterium in the ion source. Then the obtained ions are extracted from there and accelerated trough the acceleration tube to finally impact over a target after been deviated by a magnet. The target material can be either tritium or deuterium.

According to the manufacturer's specifications at 250 KV it is capable of producing a fast neutrons flux of 10^{12} neutrons/second with an energy range of up to 14 MeV.

Through out the time of our neutron generator exploitation it's been strengthening the need for an alternative way of producing ions. It is well known that there are important experiences in the world, which apply the rf power in order to ionize gases (Bernhard Wolf, 1995). Therefore we need for an rf signal to transfer to the gas contained in a volume, enough energy so as to have its particles excited to reach their ionization and to sustain the plasma.

Materials and Methods

A wide spectrum of developments and works are been performed but we will refer only to those associated with the rf application.

A device responding for generation and multiple amplification of rf signal is been upgraded and adjusted. See Figure 8.1, which pictures the structure of such a system.

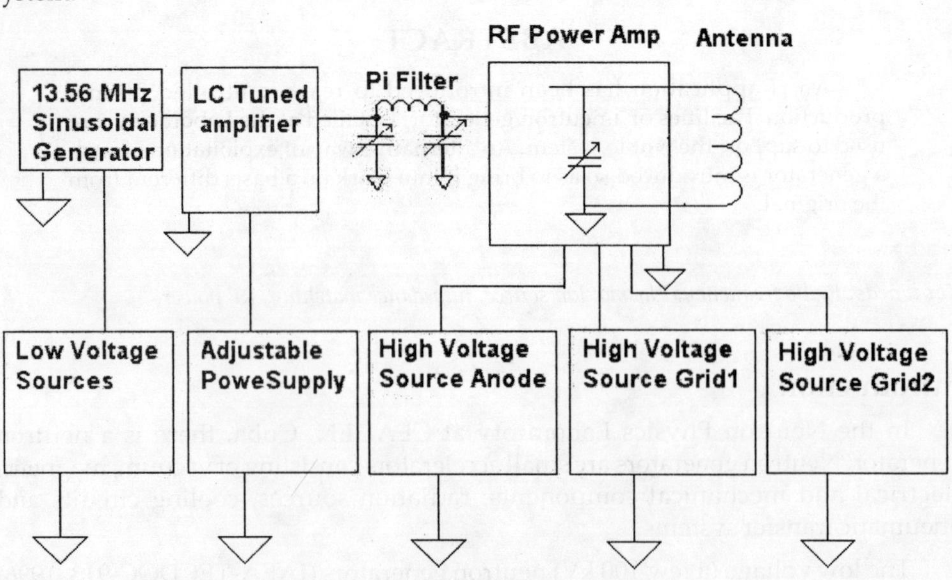

Figure 8.1: RF Amplifying System

According to the international agreements on frequency spectrum usage the 13.56 MHz channel is world wide utilized for research purposes. In our application this is the signal range.

The system is mainly based on a 13.56 MHz crystal generator and consists of multiple amplification stages, matching networks, high voltage power supplies and an antenna, which is finally located inside the ion source.

The heart of the rf power amplification track is a 5000 w rf power tube GS-9, its performance involves the use of a robust 1500 volt power supply for anode circuit polarization, built at our Lab as well as other sources that were designed and constructed to guarantee the bias of tube grids. The working point for this valve was selected so as to ensure a moderated regime of performance what is a starting point considering that it may happen that more power is required to obtain the needed output.

The signal supplied to the power stage is a sinusoidal one and it is previously amplified in an earlier stage built on a single MOSFET power transistor IRF 830. The utilized configuration is the tuned LC one as it can be seen below in the Figure 8.2. It is useful when the operating frequency is confined to a narrow range and guarantees a signal free of distortions with no harmonics (Horowitz, 1989).

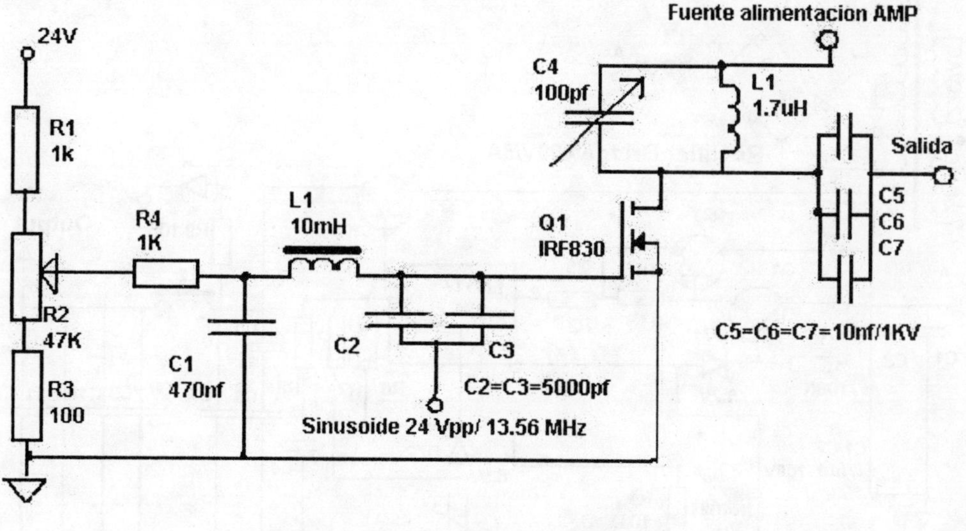

Figure 8.2: Tuned LC Amplifier

This has several advantages:

- Higher single stage gain, since the load presents high impedance at the signal frequency while allowing arbitrary quiescent current.
- Elimination of the undesirable loading effects of capacitance, since the LC "tunes out" any capacitance by making it part of the tuned circuit capacitance.

- Simplified interstage coupling, since an LC circuit can be tapped or transformer –coupled to achieve any desired impedance transformation.
- Elimination of out-of-band signals and noise owing to the frequency selectivity of the tuned circuits.

The stage recives a 24Vpp signal at the input giving out a 200Vpp sinusoidal signal. It should be commented that there is inserted between this amplification stage and the next a so called Pi filter, see Figure 8.1 in order to have the outcoming and incoming impedances matched so as to ensure the less significant rf energy losses.

At this point it was necessary to design and construct a variable power supply. See it in the Figure 8.3. It meets the following requirements:

- Variable from 1.25 V to 140 volts. It allows the control of the system output power.
- At maximum output voltage it supplies up to 300 mA. The circuit is over current protected.
- The circuit features soft start and short circuit protection.
- The dc source is rf protected so as to have no rf noise at its output level.

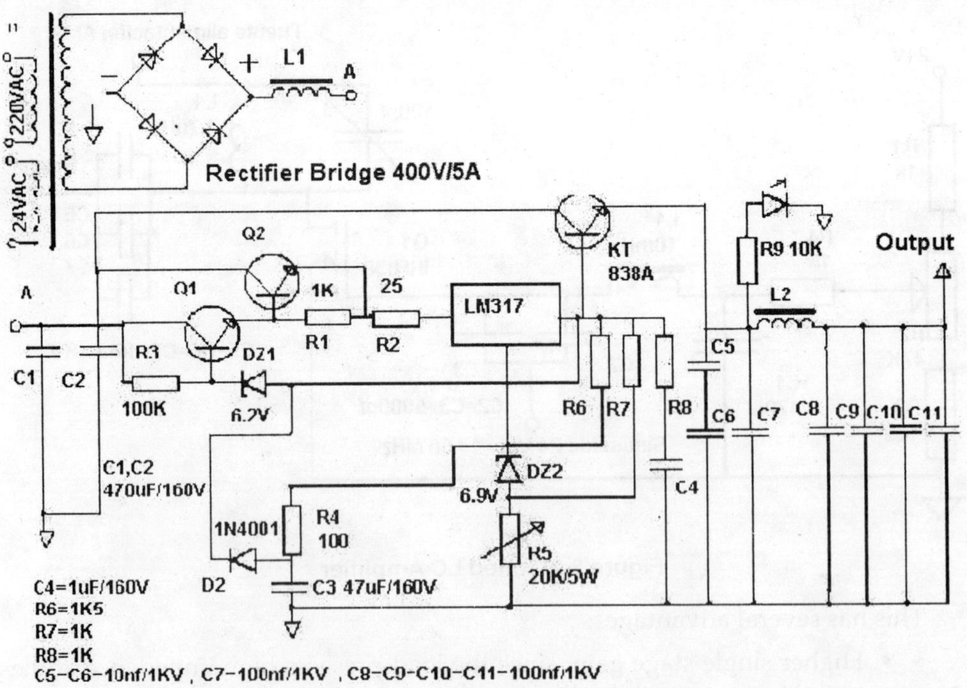

Figure 8.3: Adjustable Power Supply

It was built on discrete components. The rectifier circuit allows achieving the mayor efficiency in the alternate to direct current conversion.

The possibility of utilize the LM 117 family regulators in the floating mode makes them ideal for the operation at high voltages.

The regulator sees only the differential voltage between its input and output, which cannot exceed 40 V what is guaranteed by the 6.2V zener all over the output range from 1.25 up to 140V.

Since high voltage transistors by necessity feature low beta a Darlington pair (Q1, Q2) is used to support high voltage. Zener's impedance is low enough so there is no need for connecting capacitors directly to the regulator's input.

To limit the short circuit current to 100mA there is the 25 ohms resistance. The RC circuits at the output and at the adjusting pin improve the transient response while the 100 ohms resistance and the diode 1N4001 protects the adjusting pin during the shorts.

Q2, Q3 are placed on a radiator the last one connected as an emitter follower increases the current output of the source.

Antenna deserves a separate treatment. Considering that it is the inductor of the oscillating LC circuit at the end of the rf track it was calculated so as to achieve the resonance at the working frequency of the output LC circuit in the middle of the adjusting range. Since the environment it will be surrounded by is plasma its coating requires to be stable at high temperatures, and should be a non conductive substance. It is also required to evacuate heat from the working volume of the ion source to avoid too high temperatures and the inside of the antenna is part of the path followed by an oil flux responsible for temperature extracting.

Results

There have been carried out a number of tests during which a gas confined in a volume at low pressure was ionized due to an rf field and plasma was obtained too, see Figure 8.4.

Figure 8.4: Plasma Obtained in a Test Session

The system is very sensitive to the impedance matching and tuning. It was possible to perform coarse measurements of the rf power generated by the system and we can say that it ranges the hundreds of watts. A fine and careful adjustment is the key to diminish the losses and increase the transmitted power.

The rf power tube in the power amplifier is not working at the top of its capabilities so the working point can be readjusted and it means that there is still a source of extra energy in case it is necessary to have more of it.

Even though the system have not been tested yet in conditions closer to those it is expected to be the strength of the rf field shows an obvious ionizing capability and the maximum output is reached with no difficulties.

Conclusions

A number of devices were designed, modified, constructed to bring into work an rf system capable of supporting an rf field of medium strength to produce plasma.

Plasma production was observed. See Figure 8.4.

References

- Bernhard Wolf, (1995). Handbook of Ion Sources.
- Horowitz P., Winfield Hill. Second Edition, (1989). The Art of Electronics.
- IAEA-TECDOC-913, (1996). Manual for troubleshooting and upgrading of neutron generators.

9

Integrating RF Passives in Digital CMOS

John Richard E. Hizon[1*], Marc D. Rosales[1], Honee Lynn B. Tan[2], Louis P. Alarcon[1***] and Delfin Jay Sabido IX[3]**

[1]EEE Building Velasquez St. U.P. Diliman, Quezon City, Philippines 1101

Tel #: +632-925-2958 & Fax #: +632-925-2957

**E-mail: jehizon@up.edu.ph, chrd@ieee.org*

***E-mail: mdrosales@up.edu.ph, ***Email: lpalarcon@up.edu.ph*

[2]Gateway Business Park General Trias, Cavite

Cel #: +63917-531-4135, Email: honee.b.tan@intel.com

[3]#301 Plaza B, 6530 Northgate Ave. Alabang, Muntinlupa City, Philippines 1770

Email: jayix@standfordalumni.org

ABSTRACT

CMOS is considered the technology of choice in the complete integration of Systems on a Chip. This is attributed to the continued scaling of CMOS and its mature manufacturing technology leading to lower costs. One such application is the integration of RF transceivers in CMOS. However, CMOS is optimized for digital applications and thus presents some issues that must be addressed for the successful integration of RF in CMOS. In this article, research conducted at the Intel Microprocessors Laboratory and the Microelectronics and Microprocessors Laboratory into the integration of RF functionality in a 0.25 mm Epitaxial CMOS process are outlined. The paper would focus on the different techniques characterized in the laboratories in integrating spiral inductors and capacitors in CMOS for RF operation. These techniques include the use of Q enhancement techniques for inductors and the different capacitor topologies available in CMOS. Isolation between inductors are also considered and techniques are given on how isolation between.

Keywords: RF CMOS, PGS, Halo Substrate contacts, Fractal capacitors, Woven capacitors, Interdigital capacitors.

Introduction

Mobility requirements have propelled a lot of innovations in wireless technology. The consumer's need to be always connected anytime, has led to an explosive demand for wireless-capable devices that are small, light weight and relatively cheap. This is clearly manifested with the widespread use cellular phones. As we continue to push the limit in mobile applications, there is a need to develop devices with lower power consumption and at the same time offer acceptable processing power. These requirements can only be addressed through single chip integration. Integration offers: (1) lower cost due to a reduction in the number of external components, (2) power reduction since the need to drive large parasitic inductances and capacitances since bond pads are reduced, (3) decrease in size due to lower PCB implementations, and 4) greater functionality.

A technology of choice for system integration on chip is CMOS. With decreasing feature sizes, transistor cut-off frequency in CMOS is well within the RF domain. CMOS has other advantages namely: (1) low cost, (2) high integration density, (3) mixed signal compatibility, (4) low voltage capability, and (5) production maturity (Enz and Cheng, 2000), (Jen *et al.*, 1999), (Kim *et al.*, 1997), (Kolding, 1998), (Lee *et al.*, 2001), (Manku, 1999), (Sakalas *et al.*, 2002). However, CMOS as a technology has been optimized for digital applications. Conductive substrates and Al interconnects present limitations in the realizable Qs of inductors and capacitors while MOSFET models (BSIM3v3) used for simulations do not include relevant parasitics at RF. These limitations present roadblocks in the successful integration of RF circuits in CMOS.

RF Passives in CMOS

Research on both active (transistors) and passive (inductors, capacitors, and resistors) devices is important in realizing the ultimate goal of single-chip RF CMOS design. There is a need to determine the realizable Qs of inductors in CMOS and to evaluate different Q enhancement techniques in improving quality factor. There is also a need to determine the best capacitor implementation for RF applications to improve circuit performance. In this paper, inductors and capacitors were characterized in a 0.25 mm digital CMOS process to enable RF circuit integration. The main goal in targeting digital CMOS is for the complete integration of both baseband and RF in a common process to achieve the hard targets imposed by modern mobile applications.

Spiral Inductors

Several inductor implementations have been proposed in integrating inductors in CMOS (Mohan, 1999). Due to manufacturing issues however, only square and octagonal spiral inductors are considered in this research. Parameters that are used in designing inductors include the number of turns N, the metal trace width W, the spacing between metal traces S, the inner diameter d_{in}, outer diameter d_{out}, the average diameter $d_{ave} = 0.5(d_{out} + d_{in})$, or the fill ratio $r = (d_{out} - d_{in})/(d_{out} + d_{in})$. These parameters have been used in developing empirical equations to determine the effective inductance of spiral structures thereby simplifying its design (Mohan, 1991). However, these empirical equations do not predict the effective quality factor, Q, of a given inductor.

In the design of inductors, high Qs are necessary as it affects the performance RF blocks like LNAs and mixers (Ashby, 1996). Higher Qs are achieved when inductor losses are minimized. Losses associated with CMOS are generally classified into two: metal losses and substrate losses. Metal losses limit the realizable quality factor of inductors at lower frequencies, while substrate losses limit the quality factor at higher frequencies.

Q-enhancement techniques are used to improve inductor Q by minimizing inductor losses. Q enhancement techniques considered are patterned ground shields (Yue, 1998), halo substrate contacts (Soyuer, 1997) and shunted metal layers (Soyuer, 1995). These structures are ideal for standard CMOS processes and do not add to the cost of manufacturing the IC (Burghartz, 1997).

An inductor implemented in CMOS will typically have an effective capacitance in the order of several picofarads with the substrate (Pun, 1997), which greatly contributes to inductor induced noise at higher frequencies. Experiments on inductor induced noise were conducted for an epitaxial CMOS process (Pun, 1998), (Yeung, 1997). It was determined that increasing the distance between the noise source and the noise sensor does not reduce coupling. Studies were also conducted to study coupling between adjacent inductors (Burghartz, 1998), (Kim, 2000). It was verified that inductor coupling was a function of the distance between adjacent inductors for a lightly doped substrate. It was also verified that increasing the substrate resistivity reduces the amount of coupling between inductors for a given distance. Previous experiments on the coupling between inductors dealt with substrate coupling but the magnetic coupling between the inductors was not considered. It has been argued in (Yim, 2002) that if the coupling between inductors is by magnetic coupling, isolation characteristics between inductors will not be affected even if isolation are used. It was reported in (Yue, 1998) that using PGS contributed as much as 30 dB of isolation between adjacent inductors.

Capacitors

Capacitors are essential components in many ICs such as A/D and D/A converters, sample and hold circuits, continuous-time filters and switched-capacitor circuits. They have significant roles in RF building blocks such as low noise amplifiers (LNAs), oscillators and mixers. The capacitors in these applications consume a large portion of the chip area, thus, it is therefore desirable to have area-efficient, high density (capacitance per unit area) and high Q capacitors.

Four types of capacitors commonly used in IC design are:

1. Standard parallel plate capacitors such as metal-to-poly and metal-to-metal capacitors
2. Thin-insulator capacitors such as the poly-to-poly (double-poly) and metal-insulator-metal (MIM) capacitors.
3. Gate capacitors, also known as metal-oxide-semiconductor (MOS) capacitors
4. Junction capacitors or varactors.

The first two types of monolithic capacitors, standard parallel plate and Thin-insulator capacitors, are considered as passive capacitors. Although MIM capacitors are passive capacitors, these structures are not included in this paper primarily because their construction is not allowed in the standard digital process. Consequently, only the first type of capacitor (standard parallel plate capacitor) was investigated, together with novel structures that exploit both the vertical and lateral flux capacitances of the different metal layers. Recently developed structures that have higher capacitance densities include the interdigital, woven and quasi-fractal capacitors (Aparicio and Hajimiri, 2002), (Samavati *et al*, 1998).

Implmentation

Spiral Inductors

Several inductors were implemented for characterization. Shown in Table 9.1 are the different parameters that characterize the inductors used. Samples of the inductors implemented are shown in Figure 9.1. Q enhancement techniques like patterned ground shields, halo substrate contacts and shunted metals were then applied to these inductors to evaluate how these structures improved Q.

Table 9.1: Specification of Implemented Inductors

Inductance	*Dimensions (mm)*			
L (nH)	*N*	*W*	*S*	d_{out}
1.8 (Sq.)	3.5	10.88	4	160
1.8 (Sq.)	3.5	10.5	2.5	150
2.7 (Sq.)	3.5	10.1	3	180
3 (Oct.)	3.5	7.5	4	180
10 (Sq.)	5.5	9	4	275

The chosen structures to investigate inductor coupling experiments would answer questions on how magnetic coupling limit the isolation between inductors and on how effective isolation structures are in lessening the interaction between inductors. Isolation structures considered in this study are patterned ground shields and halo substrate contacts. These structures reduce the injection of substrate noise thus limiting substrate coupling. There were two layout configurations considered in this study. The horizontal configuration shown in Figure 9.2 enhances magnetic

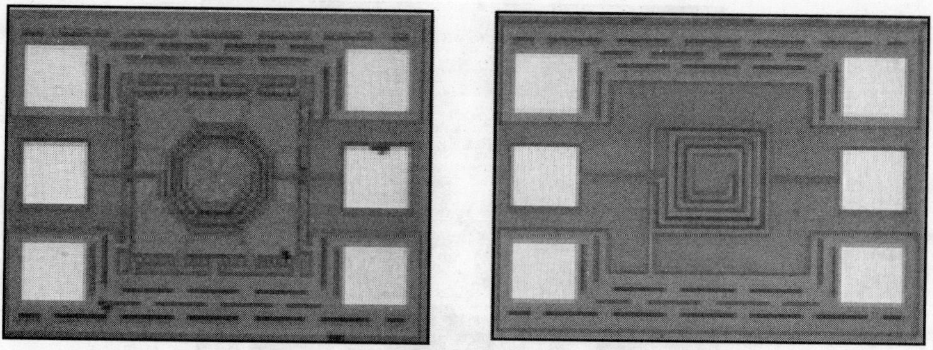

(a) (b)

Figure 9.1: Layout Sample of Inductors Implemented (a) 3 nH Octagonal Spiral Inductor with PGS and (b) 1.8 nH Spiral Inductor with Halo Substrate contacts

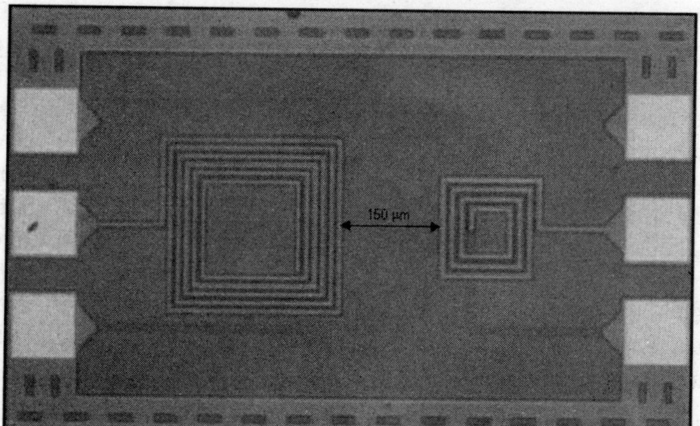

Figure 9.2: Inductors in Horizontal Configuration

coupling between inductors. On the other hand, a diagonal configuration shown in Figure 9.3 deters magnetic coupling between inductors. It can be observed that the distance between the inductors is different for the two configurations. Initially, only one distance was considered for both configurations. However, during layout construction a distance of 100 mm between inductors in a horizontal configuration was not enough to implement patterned ground shields. Thus the distance was increased to 150 mm. From previous studies, coupling in an epitaxial process is a weak function of distance when the distance considered is more than four times the thickness of the epi-layer. Thus, the distances presented do not present a significant problem in determining isolation. Note that, the inner ports of the inductors used in the coupling experiments are connected to ground. Note also that each inductor and inductor pair was placed in a test fixture for open de-embedding.

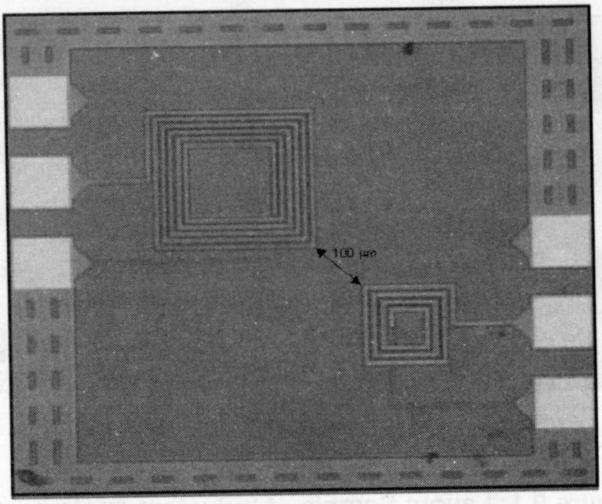

Figure 9.3: Inductors in Diagonal Configuration

Capacitors

There are various ways to construct integrated capacitors. The poly-to-metal and metal-to-metal capacitors have very good linearity and high Q but have low capacitance density. The low capacitance value is mainly due to the vertical spacing (t_{ox}) that determines the capacitance of the parallel plate structure. In today's process technologies, the vertical spacing tends to stay constant while the lateral spacing of metal layers shrinks with scaling (Aparicio and Hajimiri, 2002), (Samavati, 1998). Therefore, a higher capacitance density is achieved by taking advantage of the lateral flux capacitance together with the vertical flux capacitance. This technique is incorporated in the interdigital capacitor, the woven structure and the fractal capacitor. These three capacitor structures also have linearity similar to parallel plate capacitors (Aparicio and Hajimiri, 2002).

Lateral flux can be used to increase the total capacitance obtained in a given area since lateral spacing decreases. As a result, a higher capacitance density can be achieved by exploiting the lateral flux for each metal layer. There are various structures making use of the lateral field flux. The Vertical Parallel Plates (VPP) capacitor in Figure 9.4 is constructed using multiple vias that vertically connect the metal strips of each layer. The Vertical Bars (VB) capacitor structure is an improved modification of the VPP capacitor. This capacitor has a higher capacitance density because it takes advantage of the capacitance along the x-axis (c_x) and y-axis (c_y).

The Fractal capacitor on the other hand, may be implemented using purely lateral flux or both the lateral flux and the vertical flux. Two notable advantages in using fractal capacitors [Samavati, 1998] in the construction of a monolithic capacitor:

1. Higher density of fractal capacitor results in a smaller area when both the lateral flux and vertical flux are utilized.
2. Parasitic bottom plate capacitance converted to useful plate-to-plate capacitance because some of the field lines from the bottom plates terminate on the adjacent plate instead on the substrate.

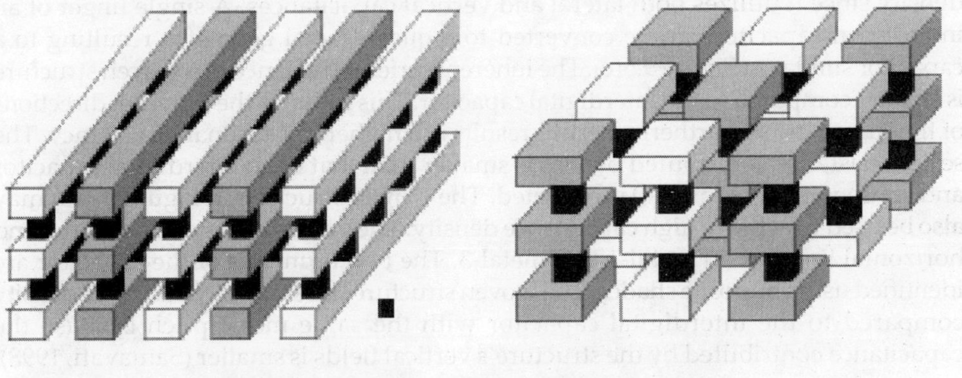

(a) (b)

Figure 9.4: Lateral Flux Capacitors
(a) Vertical Parallel Plates and (b) Vertical Bars

However, problems may be encountered in the fabrication of fractal capacitor such as lithography limitations. A solution to these problems is the use of Quasi-Fractal geometry [Aparicio & Hajimiri, 2002]. Figure 9.5(a) is an example of a Quasi-Fractal capacitor that maximizes the use of purely lateral flux. The capacitor structures in Figures 5(b) and 5(c) have higher capacitance densities by using both the vertical and lateral fields [Aparicio & Hajimiri, 2002]. These Quasi-Fractal capacitors, compared to HPP capacitors, have higher capacitance density and the same linearity [Aparicio & Hajimiri, 2002].

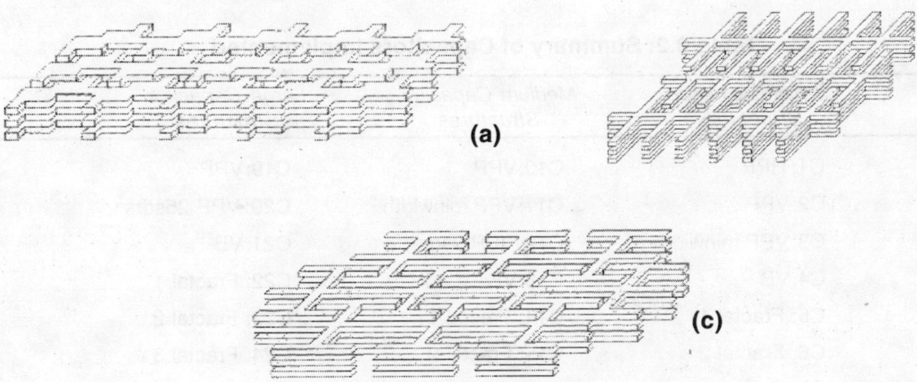

(a) (b) (c)

Figure 9.5: Quasi Fractal Capacitors
(a) Fractal 1 (b) Fractal 2 and (c) Fractal 3

The last two structures, Interdigital and Woven capacitors, achieve a higher capacitance density by combining the lateral flux and the vertical flux. The Interdigital capacitor, shown in Figure 9.6(a), is another simple way of increasing capacitance density since it utilizes both lateral and vertical capacitances. A single finger of an interdigital capacitor can be converted to a quasi-fractal geometry resulting to a capacitor similar to Figure 9.5(c). The inherent series inductance of a woven structure is smaller compared to the interdigital capacitor. This is due to the different directions of the current flow. Furthermore, this results in a higher self-resonant frequency. The series resistance contributed by vias is smaller than that of an interdigital capacitor and consequently, a large Q is expected. The woven structure in Figure 9.6(b) may also be used to achieve high capacitance density. The vertical lines are in metal-2 and horizontal lines are in metal-1 and metal-3. The two terminals of the capacitor are identified using opposite shades. The woven structure has smaller capacitance density compared to the interdigital capacitor with the same metal pitch because the capacitance contributed by the structure's vertical fields is smaller (Samavati, 1998).

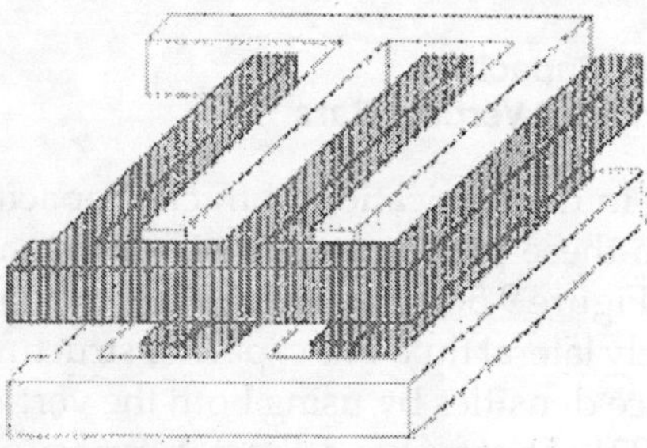

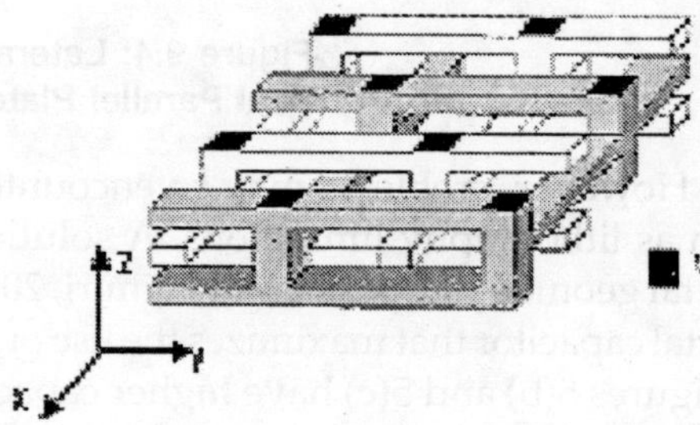

Figure 9.6: Vertical and Lateral Flux Capacitors (a) Interdigital and (b) Woven

Table 9.2: Summary of Capacitors Implemented

Small Capacitor Structures	*Medium Capacitor Structures*	*Large Capacitor Structures*
C1: HPP	C10: VPP	C19: VPP
C2: VPP	C11: VPP minwidth	C20: VPP 25sqrs
C3: VPP minwidth	C12: VPP 9sqrs	C21: VB
C4: VB	C13: VB	C22: Fractal 1
C5: Fractal 1	C14: Fractal 1	C23: Fractal 2
C6: Fractal 2	C15: Fractal 2	C24: Fractal 3
C7: Fractal 3	C16: Fractal 3	C25: Interdigital
C8: Interdigital	C17: Interdigital	C26: Woven
C9: Woven	C18: Woven	

Only a small Horizontal Parallel Plate (HPP) capacitor was implemented because this type of structure occupies a large portion of the die due to its size. The rest of the structures, specifically the Vertical Parallel Plates (VPP), Vertical Bars (VB), Quasi-

Fractal, Interdigital and Woven capacitors, were constructed in three major groups. - The first group is the small capacitances with values near 600 fF; the second group having capacitances of around 3 pF to 6 pF; and the third group is the large capacitors with values ranging from 6 pF to 12 pF. Table 9.2 gives a short listing of all the capacitors implemented.

Measurement and Modeling

The measurement set-up used in this study is shown in Figure 9.7. On-wafer measurements were conducted using a Micromanipulator Probing station, Microwave Probe Link Arms, GSG Picoprobes 40A-GSG-160-P, an Agilent 8753ES Vector Network Analyzer, a HP 4156A Precision Semiconductor Parameter Analyzer, and HP 11590B bias tees. All test structures were enclosed in test fixtures for de-embedding processes. Open pad de-embedding and three step de-embedding structures were included in all fabricated chips and it was determined that open pad de-embedding was sufficient for frequencies up to 6 GHz.

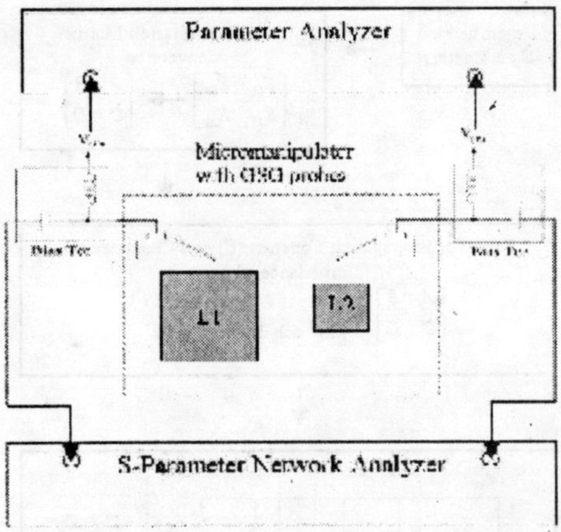

Figure 9.7: Measurement Set up Used

Spiral Inductors

The model used to characterize implemented inductors was presented by Yue. The inductor model was derived using the flowchart illustrated in Figure 9.8. Each component value is computed per frequency point with Zo = 50 ohms. Note that Cs was not determined from measured values since the self-resonant frequency of the branch containing Cs is beyond 6 GHz.

There were 9 different bias conditions considered in this study as shown in Table 9.3. The bias conditions specified in cases 2 to 5 determine if the presence of a DC current affects the amount of coupling between adjacent inductors. These cases also determine if the amount of coupling is dependent on whether the DC current flows in a large inductor or a small inductor. Bias Conditions specified in cases 6 to 9 determine if inductor coupling is affected by the polarity of the mutual magnetic coupling between adjacent inductors.

Table 9.3: Different Bias Condition Considered in the Experiment

Bias Condition	*Large (10 nH)*	*Small (1.8 nH)*
1	no bias	no bias
2	+	no bias
3	–	no bias
4	no bias	+
5	no bias	–
6	+	+
7	+	–
8	–	+
9	–	–

LEGEND:
+: Current flowing into the inductor; –: Current flowing out of the inductor.

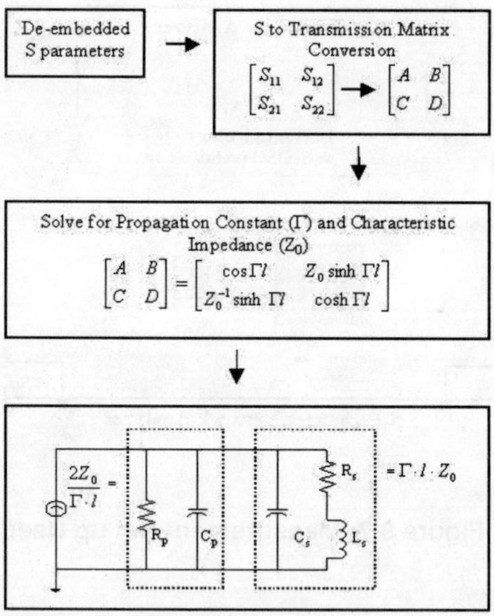

Figure 9.8. Parameter Extraction by Yue

From measured s-parameters, the coupling factor as presented in [Ng, 2002] is extracted to determine the contribution of substrate and magnetic coupling on inductor coupling. The mutual reactive coupling factor, k_{Im}, is determined using the expression given in Equation (1) and the mutual resistive coupling factor, k_{Re}, is given by Equation (2). Note that these expressions are used in studying spiral transformers but were used in this study since adjacent inductors behave like weak transformers.

$$k_{\mathrm{Im}} = \sqrt{\frac{[\mathrm{Im}(Z_{12})]^2}{\mathrm{Im}(Z_{11})\mathrm{Im}(Z_{22})}} \qquad (1)$$

$$k_{Re} = \sqrt{\frac{[Re(Z_{12})]^2}{Re(Z_{11})Re(Z_{22})}} \quad (2)$$

Capacitors

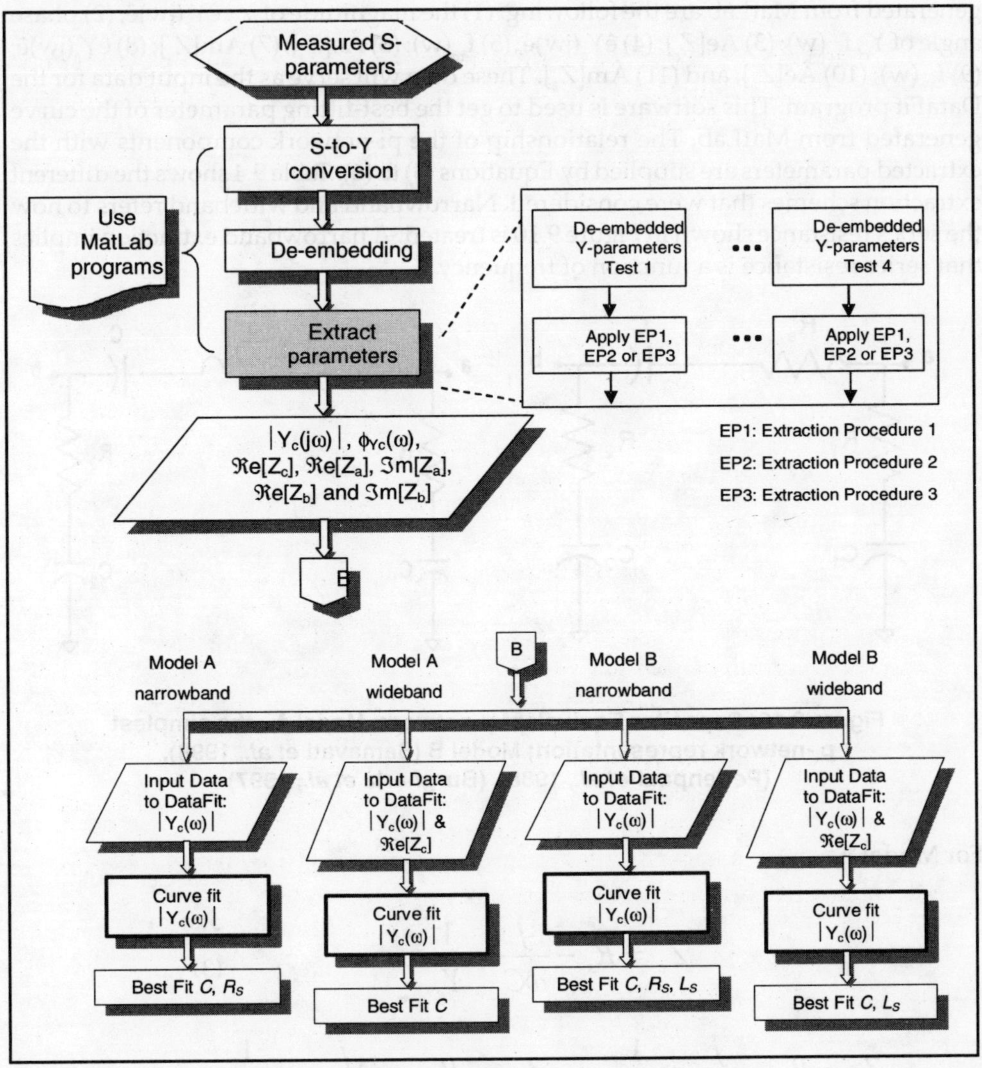

Figure 9.9: Extraction Procedure Flow Chart

The flow chart of the extraction procedure is shown in Figure 9.9. The elements of the equivalent pi-network model can be obtained by further manipulating the measured S-parameters. These elements are: (1) component capacitance, C; (2) effective series resistance, R_s; (3) inherent series inductance, L_s; (4) substrate parasitic capacitances (C_1 and C_2) of the respective ports 1 and 2; and (5) substrate parasitic resistances (R_1 and R_2) of the respective ports 1 and 2. The extraction is performed by first converting the S-parameters to Y-parameters, then the two de-embedding techniques are applied. These de-embedded Y-parameters are applied to a specific extraction procedure. The parameter extraction and data manipulation is done using the MatLab scripts. The parameters extracted from the de-embedded Y-parameters generated from MatLab are the following: (1) the magnitude of Y_c, êY_c(jw)ê; (2) phase angle of Y_c, f$_{Yc}$(w); (3) Âe[Z_c]; (4) êY_a(jw)ê; (5) f$_{Ya}$(w); (6) Âe[Z_a]; (7) Ám[Z_a]; (8) êY_b(jw)ê; (9) f$_{Yb}$(w); (10) Âe[Z_b]; and (11) Ám[Z_b]. These data will serve as the input data for the DataFit program. This software is used to get the best-fitting parameter of the curve generated from MatLab. The relationship of the pi-network components with the extracted parameters are supplied by Equations (3) to (6). Table 9.4 shows the different extraction schemes that were considered. Narrowband and wideband refers to how the series resistance shown in Figure 9.10 is treated. A narrowband extraction implies that series resistance is a function of frequency.

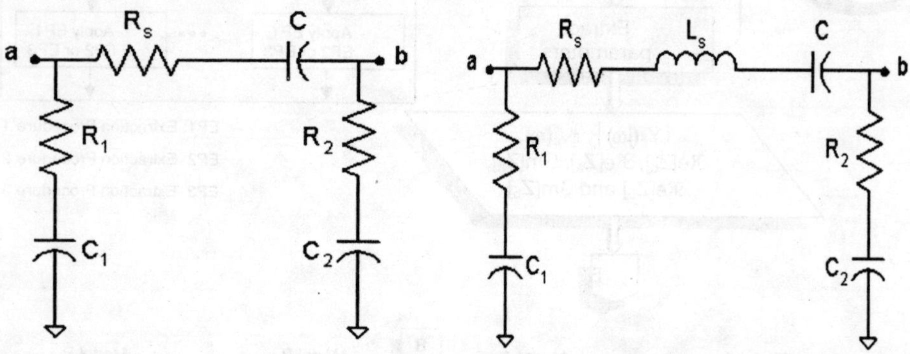

Figure 9.10: Capacitor Equivalent p-network: Model A - the simplest p -network representation; Model B (Samavati *et al.*, 1998), (Pettenpaul *et al.*, 1988), (Burghartz *et al.*, 1997)

For Model A

$$Z_c = R_s - \frac{j}{\omega C} = \frac{1}{Y_c} \qquad (3)$$

$$Z_a = R_1 - \frac{j}{\omega C_1} = \frac{1}{Y_a} \text{ and } Z_b = R_2 - \frac{j}{\omega C_2} = \frac{1}{Y_b} \qquad (4)$$

For Model B

$$Z_c = R_s + j\left(\omega L_s + \frac{1}{\omega C}\right) = \frac{1}{Y_c} \qquad (5)$$

$$Z_a = R_1 - \frac{j}{\omega C_1} = \frac{1}{Y_a} \text{ and } Z_b = R_2 - \frac{j}{\omega C_2} = \frac{1}{Y_b} \qquad (6)$$

Another parameter that has to be extracted is the quality factor, *Q*. Equation (7) is the equation of *Q* that represent the reactance and effective series resistance (*ESR*) of the impedance across the two terminals. The Q solved from the de-embedded Y_{12} parameter is defined as the component Q ($Q_{component}$). Another approach on the extraction of the Q is to get the circuit Q ($Q_{circuit}$). The circuit Q is achieved by considering the effect of the parasitics contributed by the substrate. This is done by connecting port 2 for any model given in Figure 9.10 to ground and get the effective admittance as seen at port 1. The result is a parallel combination of Y_c and Y_a. The effective admittance is given by Equation (8) and the circuit Q by Equation (9).

Table 9.4. Extraction Schemes used in Characterizing Inductors

Scheme	*OPD*	*TRISD*	*Model A*	*Model B*	*Narrowband*	*Wideband*
1	✓		✓		✓	
2	✓		✓			✓
3	✓			✓	✓	
4	✓			✓		✓
5		✓	✓		✓	
6		✓	✓			✓
7		✓		✓	✓	
8		✓		✓		✓

$$Q_{component} = \frac{-\operatorname{Im}(Z_c)}{\operatorname{Re}(Z_c)} = \frac{\operatorname{Im}\left(\frac{1}{-Y_{12,de-embedded}}\right)}{\operatorname{Re}\left(\frac{1}{-Y_{12,de-embedded}}\right)} = \frac{\operatorname{Im}(Y_{12,de-embedded})}{\operatorname{Re}(Y_{12,de-embedded})} \qquad (7)$$

$$Y_d = Y_{11} + Y_{12} - Y_{12} = Y_{11} \qquad (8)$$

$$Q_{circuit} = \frac{\operatorname{Im}\left(\frac{1}{Y_{11,de-embedded}}\right)}{\operatorname{Re}\left(\frac{1}{-Y_{11,de-embedded}}\right)} = \frac{\operatorname{Im}(Y_{11,de-embedded})}{\operatorname{Re}(Y_{11,de-embedded})} \qquad (9)$$

Discussion of Results

For small inductors, the use of shunted metals has the greatest contributor in increasing Q for frequencies below 3 GHz. However, when frequency is increased above 3 GHz the substrate losses begin to dominate making the shunted metal technique ineffective. Looking at Figure 9.11(a), for frequencies above 4 GHz, using a plain inductor offers the best possible quality factor.

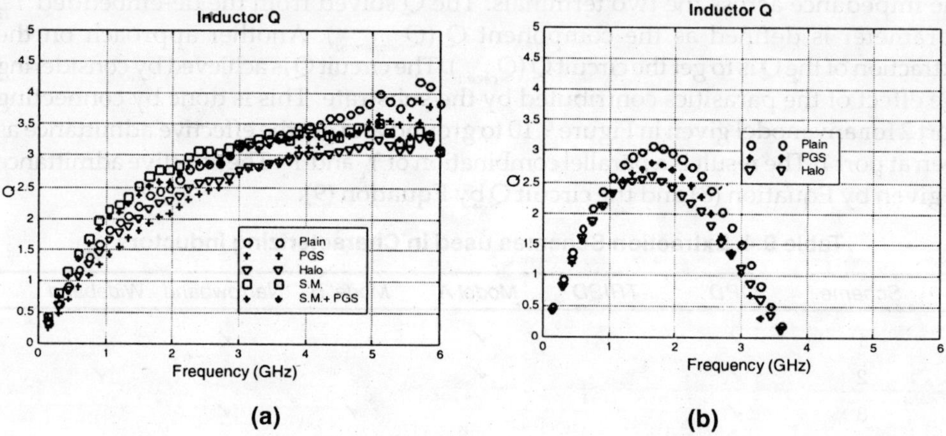

Figure 9.11: Extracted Q: (a) 1.8 nH (b) 10 nH

The quality factor extracted from 10 nH inductors using patterned ground shields is better when compared with inductors using halo substrate contacts as seen in Figure 9.11(b). However, the high shunt capacitance associated with patterned ground shields reduces the inductors resonant frequency. It can be verified from the Figure 9.11 that the use of plain inductors gives the highest attainable quality factor.

It can be observed that ocatagonal spiral inductors with Q enhancemnets have better Qs compared with square spiral inductors also with Q enhancements. Shown in Figure 9.12 are plots comparing the Q for both topologies using PGS and shunted metal layers. As frequency increases, octagonal spirals tend to have higher Qs than their square counterparts. This is attributed to lower shunt capacitances of octagonal spirals due to their hollow implementations.

From the summary shown in Figure 9.13, it can be verified that using a diagonal configuration improves isolation by 5 dB for each of the three cases considered except for inductors with PGS when frequency exceeds 3 GHz. From the same Figure it can also be verified that using PGS helps in providing isolation between inductors as compared to halo substrate contacts. It can be observed that as frequency increases, the amount of coupling increases for inductors using these structures. The use of substrate contacts effectively provides a low impedance path from the epitaxial layer to the bulk substrate.

Diagonal Layout

From Figure 9.14(a), it can be observed that the use isolation structures does not affect the amount of electric coupling between inductors for frequencies below 2

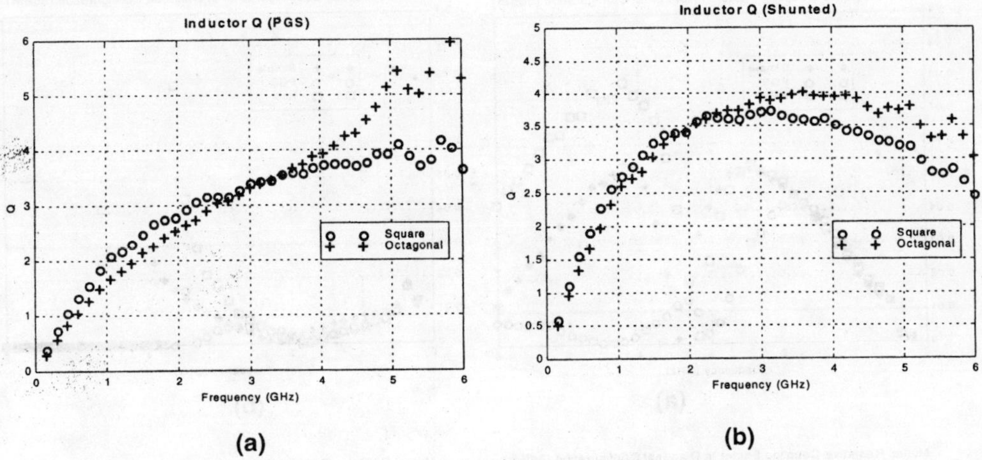

Figure 9.12: Extracted Q: (a) 3 nH with PGS (b) 3 nH with Shunted Metal Layers

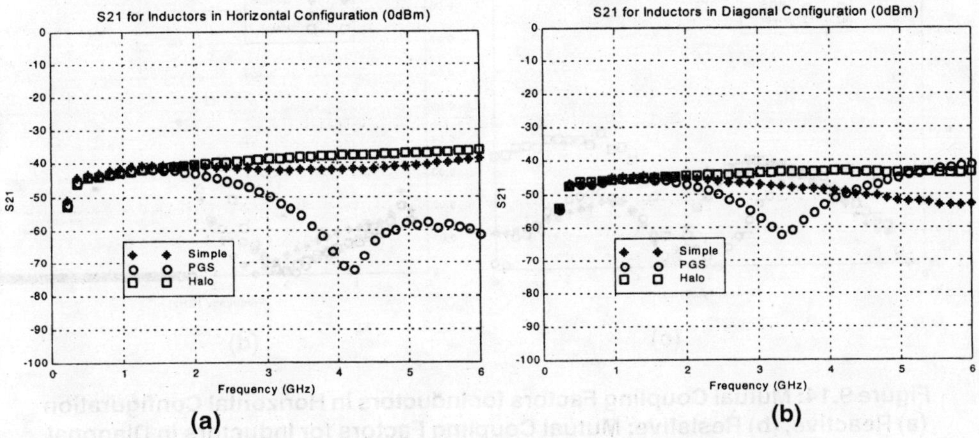

Figure 9.13: Summary of Coupling Results: (a) Horizontal Layout

GHz. But for frequencies beyond 3 GHz, it can be observed that structures with substrate contacts have the largest coupling factor verifying the results obtained in Figure 9.13. The same trend was observed for the diagonal structures as seen in Figure 9.14(c). It should be noted that there is a larger electric coupling in the horizontal configuration due to a larger parasitic capacitance between the inductors.

From Figure 9.14(b & d), it can be observed that the mutual reactive coupling factor for adjacent inductors with isolation structures is lower compared to adjacent inductors without isolation structures. The lowering of the mutual reactive coupling factor is attributed to the slightly lower inductance achieved when patterned ground shields and halo substrate contacts are used. Since the magnetic coupling between adjacent inductors in a horizontal configuration is higher, the lowering of the inductance has a more pronounced effect for the horizontal configuration.

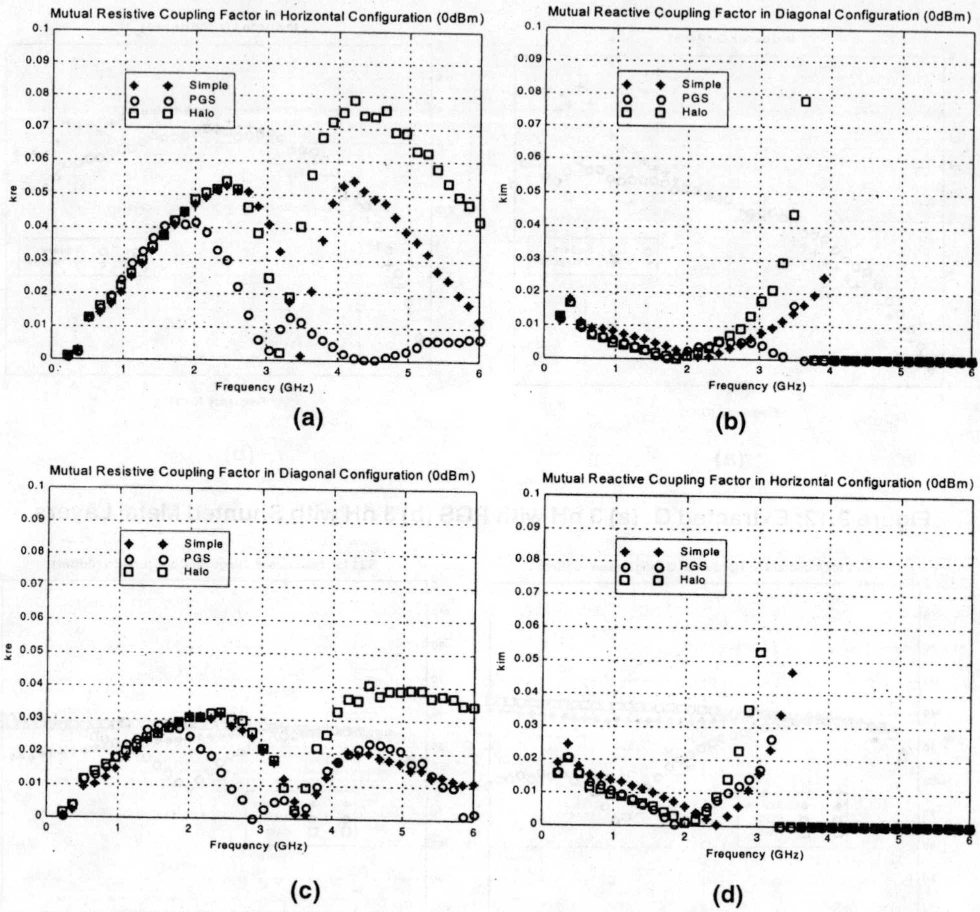

Figure 9.14: Mutual Coupling Factors for Inductors in Horizontal Configuration (a) Reactive, (b) Resistive; Mutual Coupling Factors for Inductors in Diagonal Configuration: (a) Reactive (b) Resistive

The capacitance density resulting from extraction schemes 4 and 8 is illustrated in Figure 9.15. The capacitance density of each capacitor structure is fairly constant since the increase in the capacitor value is proportional to increase in its area. The medium capacitors' area and capacitance is approximately eight times (~8x) that of the small capacitors' while the large capacitors' area and capacitance is about twice (~2x) that of the medium capacitors'. Some of the variation in the capacitance density (*e.g.* VB and Woven) is contributed by process variation in small metal squares, because with these types of structures, the fringing effect is comparable to the actual capacitance due to their small dimensions and close proximity to each other.

In Figure 9.15, the woven structures have the highest capacitance density, followed by the VPP minwidth capacitors and VPP capacitor units (VPP 9 sqrs and VPP 25 sqrs), then the vertical bar capacitors and the interdigital capacitors. The

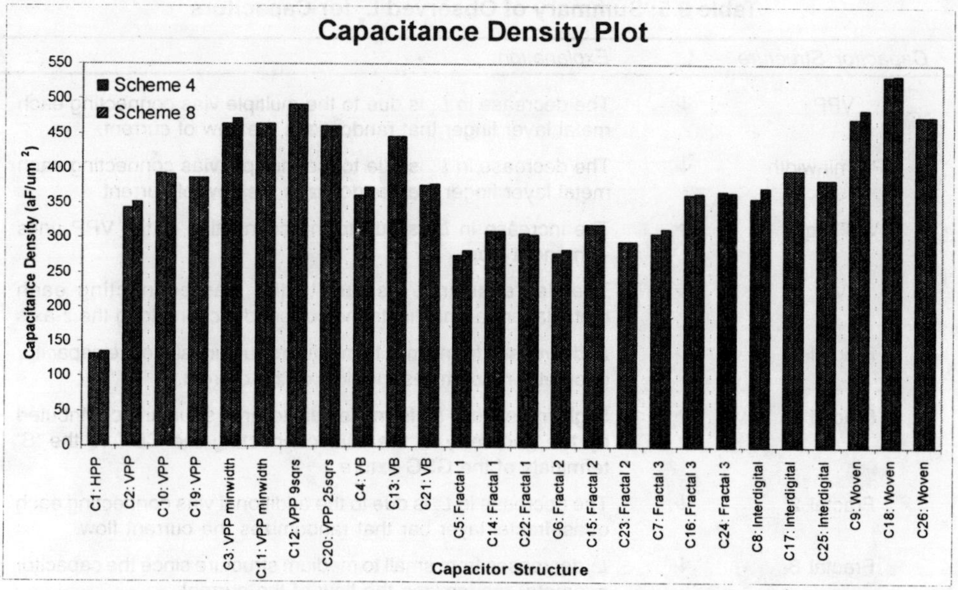

Figure 9.15: Capacitance Density (the increasing assignment of capacitor number indicate the increase in capacitor size for each structure type)

quasi-fractal structures are expected to have lower capacitance density because of the area they occupy. Figure 9.15 shows that the C1: HPP capacitor has the smallest capacitance density – a disadvantage for area constrained design. The capacitance density of C3: VPP minwidth and C9: Woven capacitors are 3.5 times the capacitance density of the C1: HPP capacitor – a 350% increase in capacitance density.

The initial estimate of the self inductance of the wire can be calculated using Equation (10). A summary of the observed result on L_S is described in Table 9.5 along with an explanation on the effect of the capacitor structure and the interconnect geometry that contributed to this parasitic. The direction of the arrows indicates the trend of L_S as the capacitor structure is increased.

$$L_{self} = \frac{\mu_0 \mu_r l}{2\pi}\left[\ln\left(\frac{2l}{w+t}\right) + 0.5 + \frac{0.447(w+t)}{2l}\right] \tag{10}$$

The resonant frequency (f_0) of each of the capacitor is also obtained from extraction scheme 4. The resonant frequency is expected to decrease as the capacitance value is increased. This assumption is always true if the series inductance does not change with capacitor size. The relationship of the resonant frequency with the series inductance and capacitance is given by Equation (11). It is preferred to have a high resonant frequency because this implies that below the resonant frequency, the structure behaves like a capacitor. All the resonant frequencies of the capacitors are above the 2.4 GHz center frequency of the LNA. These fabricated capacitors can be used in the design of the LNA operating at 2.4 GHz.

Table 9.5: Summary of Observed L_s for Capacitors

Capacitor Structure	L_S	*Explanation*
VPP	↓	The decrease in L_S is due to the multiple vias connecting each metal layer finger that randomizes the flow of current.
VPP minwidth	↓	The decrease in L_S is due to the multiple vias connecting each metal layer finger that randomizes the flow of current
VPP sqrs	↑	The increase in L_S is due to the connection of the VPP units -forming a loop.
VB	↓	The decrease in L_S is due to the vias connecting each metal layer bar that limits the current direction along the z-axis
Fractal 1	↓	L_S decreases from small to medium structure since the capacitor geometry randomizes the flow of the current.
Fractal 1	↑	Slight increase in L_S, from medium to large structure, contributed by the geometry of the wire connecting the CUT to the 'S' terminals of the GSG fixture
Fractal 2	↓	The decrease in L_S is due to the additional vias connecting each quasi-fractal layer bar that randomizes the current flow.
Fractal 3	↓	L_S decreases from small to medium structure since the capacitor geometry randomizes the flow of the current.
Fractal 3	↑	Slight increase in L_S, from medium to large structure, contributed by the geometry of the wire connecting the CUT to the 'S' terminals of the GSG fixture
Interdigital	↓	The decrease in L_S is due the small finger lengths connected alternately per metal layer.
Woven	↓	L_S decreases from small to medium structure because the size can, in effect, randomize the flow of the current
Woven	↑	L_S increases from medium to large structure because as the size is increased, the randomized effect of the current flow lessens.

$$f_0 = \frac{1}{2\pi\sqrt{CL_S}} \qquad (11)$$

The resonant frequency plot of all the capacitor structures, in Figure 9.16, shows that f_0, in general, decreases with increasing capacitance. The effect of the series inductance with f_0 and its relationship with the capacitance must also be taken into account. That is the reason why some of the structures exhibit a slight increase in the resonant frequency as the size is increased from medium to large structure. This behavior occurs when the series inductance is significantly reduced as the capacitance value is increased - a case where the change in the L_S value is not proportionately the same as the change in the capacitance. This is evident in the Fractal 2 capacitor structure.

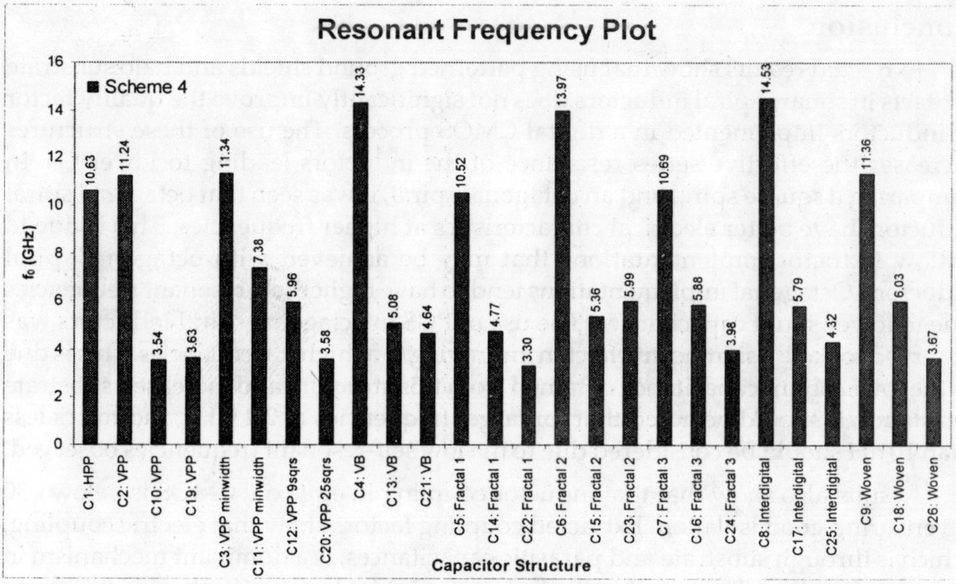

Figure 9.16: Resonant Frequency of All Capacitor Structures

The quality of each of the capacitor structure degrades with increasing frequency and capacitor size. On the contrary, the dissipation factor, *DF*, increases, thus indicating that the *ESR* becomes prominent at higher frequencies and bigger capacitances. An example of this observation in the comparison of the Q and *DF* per capacitor structure is presented in Figure 9.17.

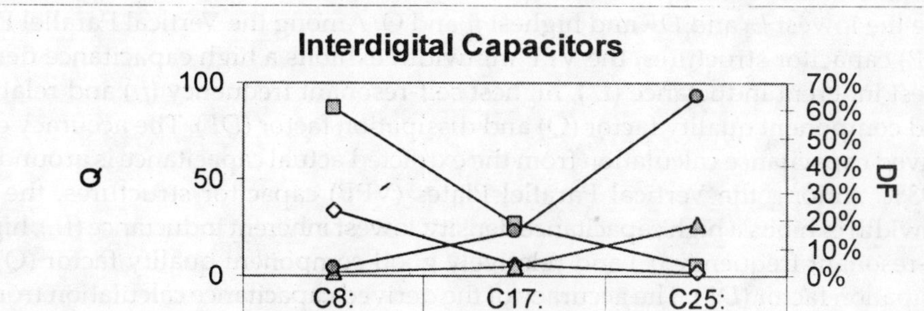

	C8: Interdigital	C17: Interdigital	C25: Interdigital
Q, 1.02GHz	87.20	26.99	5.35
Q, 2.4GHz	34.41	6.12	1.54
DF, 1.02GHz	1.15%	3.71%	18.69%
DF, 2.4GHz	2.91%	16.34%	65.07%

Capacitor Structure

Figure 9.17: Q and DF Plot for Interdigital Capacitor; the Q becomes smaller with increasing frequency and capacitor size while the opposite is true with the DF

Conclusion

Extracted results show that using patterned ground shields and halo substrate contacts in square spiral inductors does not significantly improve the quality factor of inductors implemented in a digital CMOS process. The use of these structures increases the effective series resistance of the inductors leading to lower Qs. In comparing a square spiral and an octagonal spiral, it was seen that octagonal spiral inductors have better electrical characteristics at higher frequencies. This is due to hallow inductor implementations that may be achieved with octagonal spiral inductors. Octagonal implementations tend to have higher self-resonant frequencies due to lower shunt capacitances. The use of PGS in octagonal spiral inductors was observed to have a significant effect in improving Q at higher frequencies. This is due to the lower shunt capacitance obtained from this structure and increased substrate resistance. It should be noted that for target frequencies of 2.4 GHz, inductors less than 10 nH should be considered due to the low self-resonant frequencies observed.

Results also show that that inductor coupling in epitaxial CMOS is below –30 dB ensuring good isolation. Extracted coupling factors show that electric coupling, which is through substrate and parasitic capacitances, is a dominant mechanism in inductor coupling. The use of patterned ground shields was effective in improving isolation but may be limited due to area constraints. Using a diagonal configuration limits both electric and magnetic coupling thereby contributing to isolation improvements of at least 5 dB. Results also show that isolation between inductors is not dependent on input power below 0 dB, DC current through an inductor and magnetic coupling polarity.

The three capacitor structures (VPP minwidth, Fractal 3 and VB) are recommended for RF applications operating at 2.4 GHz. The chosen three structures have the lowest L_S and *DF* and highest f_0 and Q. Among the Vertical Parallel Plates (VPP) capacitor structures, the VPP minwidth exhibits a high capacitance density, lowest inherent inductance (L_S), highest self-resonant frequency (f_0) and relatively good component quality factor (Q) and dissipation factor (*DF*). The accuracy of the derived capacitance calculation from the extracted actual capacitance is around 90% to 93%. Among the Vertical Parallel Plates (VPP) capacitor structures, the VPP minwidth exhibits a high capacitance density, lowest inherent inductance (L_S), highest self-resonant frequency (f_0) and relatively good component quality factor (Q) and dissipation factor (*DF*). The accuracy of the derived capacitance calculation from the extracted actual capacitance is around 90% to 93%. Among the quasi-fractal capacitors the Fractal 3 capacitor has the highest capacitance density, lowest L_S, highest f_0, acceptable Q and. The disadvantage, however, of the quasi-fractal structures is in its construction because the generation of the fractal segments is not computer aided.

Although the vertical bars (VB) capacitor falls under the purely lateral capacitors, the presence of the additional capacitance formed along the *z*-axis combines the effect of the vertical and lateral capacitances. Hence comparison of VB capacitors among the Interdigital and Woven structures can be considered. Among the VB, Interdigital and woven capacitors, VB structure has the lowest L_S and *DF* and highest f_0 and Q. The capacitance density is acceptable since it falls within the capacitance density range of the Interdigital and Woven structures.

Acknowledgements

We would like to thank Intel Philippines, DOST- PCASTRD, I.M.I. and Eazix Inc. for their support to the UP IML-MICROLAB of the EEE Department, U.P., Diliman.

References

- Aparicio, R. and Hajimiri A., (2002). Capacity Limits and Matching Properties of Integrated Capacitors. *IEEE J. Solid-State Circuits*, vol. 37, 384-393.
- Ashby, K.B. *et al.*, (1996). High Q inductor for wireless applications in a complementary silicon bipolar process. *IEEE J. Solid-State Circuits*, vol. 31, 4-9.
- Burghartz, J.N. *et al.*, (1998). RF design aspects of spiral inductors in silicon. *IEEE J. Solid-State Circuits*, vol. 33, No. **12:** 2028-2034.
- Burghartz, J., (1997). Silicon RF Technology-Two Generic Approaches. *Proc. ESSDERC*, 143-153.
- Burghartz, J., *et al.*, (1997). Integrated RF Components in a SiGe Bipolar Transistor. *IEEE J. Solid-State Circuits*, vol. 32, 1440-1445.
- Enz, C. and Cheng, Y. (2000). MOS Transistor Modeling for RF IC Design. *IEEE Transactions on Solid-State Circuits*, Vol. 35, No. **2**: 186-201.
- Kim, C. *et al.*, (2000). Design guide of coupling between inductors and its effect on reverse isolation of a CMOS LNA. *Tech. Digest of MTT IMS2000*.
- Kim, C. *et al.* (1997). CMOS Layout and Bias Optimization for RF IC Design Applications. *IEEE MTT-S Digest*, 945-948.
- Kolding, T. (1998). Review of RF CMOS Performance and Future Process Innovations. Technical Report R98-1014, RF Integrated Systems and Circuits (RISC) Group, Aalborg University, Denmark.
- Lee, S. *et al.*, (2001). A Small-Signal RF Model and its Parameter Extraction for Substrate Effects in RF MOSFETs. *IEEE Transactions on Electron Devices*, Vol. 48, No. **7:** 1374-1379.
- Manku, T., (1999). Mirowave CMOS Device – Device Physics and Design. *IEEE Journal of Solid-State Circuits*, Vol. 34, No. **3**: 277-85.
- Mohan, S.S. *et al.*, (1999). Simple accurate expressions for planar spiral inductors. IEEE J. Solid-State Circuits, vol. 34, 1419-1425.
- Ng, K.T. *et al.*, (2002). Substrate effects in monolithic RF transformers on silicon. *IEEE Transactions on Microwave Theory and Techniques*, vol. 50, No.**1:** 377-383.
- Pettenpaul, *et al.*, (1998). The simulation and Design of Integrated Inductors. *IEEE Trans. on Microwave Theory & Techniques*, **36(2):** 294-304.
- Pun, A. *et al.*, (1998). Substrate noise coupling through planar spiral inductor. *IEEE J. Solid-State Circuits*, vol. 33, No. **6:** 877–884.
- Pun, A. *et al.*, (1997). Experimental results and simulation of substrate noise coupling via planar spiral inductor in RF IC's. Tech. Dig. Intern. Electron Devices Meeting (IEDM), 325-328.

- Sakalas, P. *et al.*, (2002). Impact of Pad and Gate Parasitics on Small-Signal and Noise Modeling of 0.35mm Gate Length MOS Transistors. *IEEE Transactions on Electron Devices*, Vol. 49, No. **5:** 871-879.
- Samavati, H *et al.*, (1998). Fractal Capacitors. *IEEE J. Solid-State Circuits*, vol. 33, 2035-2041.
- Soyuer, M. and Nguyen-Ngoc, D., (1997). Novel substrate contact structure for high Q silicon integrated spiral inductors. *Tech. Digest International Electron Devices Meeting*, 55-58.
- Soyuer, M. *et al.*, (1995). Multi-level monolithic Inductors in Silicon Technology. *El. Letter*, vol. 31, No. **5:** 359-360.
- Yeung, T. *et al.*, (1997). Noise coupling in heavily and lightly doped substrate from planar spiral inductor. *Proc. of IEEE International Symposium on Circuits and Systems (ISCAS)*, vol. 2, 1405-1408.
- Yue, C.P. and Wong, S.S., (1998). On chip spiral inductors with patterned ground shields for Si-based RF IC's. *IEEE J. Solid-State Circuits*, vol. 33, 743-752.
- Yim, S. and O, K., (2002). The effects of a ground shield on the characteristics and performance of spiral inductors. *IEEE J. Solid-State Circuits*, vol. 37, No. **2:** 237-244.

10

Microelectronics in Turkey

Gül Sentürk
Researcher, Microelectronics Division,
National Research Institute of Electronics and Cryptology,
Gebze/Kocaeli, Turkey
E-mail: uekae@uekae.tubitak.gov.tr, gulsenturk@uekae.tubitak.gov.tr

ABSTRACT

Microelectronics Research Laboratory of TUBITAK- UEKAE called YITAL is the unique microelectronics laboratory in Turkey with its capability of giving RTD services on CMOS integrated circuits. Since its foundation in 1983, YITAL has been researching semiconductor technologies and developing production processes in a 500-m² area in TUBITAK Gebze. In the beginning YITAL was a research laboratory with 1000 class clean room. Today YITAL produces 1.5µm CMOS IC's and has capability of 0.7µm CMOS IC productions within 10-class clean room. We have the potential to build on this capability to achieve levels below half micron.

Introduction

The information and Data Processing Technologies have shown a remarkable development during the recent years. To take part among the developed countries in the 21st. century, which is called "information age" depends upon the electronic and communication sectors also becoming capable of producing technology as well as competitiveness of our products with those produced by other countries in our globalizing world.

The electronic sector is a very much promising sector to create the highest employment in the 21st Century with it's already existing about US $1 trillion market volume all around the world in general. Now the Electronics Industry is not a simple industry branch anymore, but rather turned into a basic and generic industry by itself to develope all the other industries.

However, much of the Turkish Electronic Industry may be seen far from being at desirable dimensions when viewed with respect to world-wide figures and its share in the Turkish Economy.

The total production of electronic industry in 2004 increased from 3,5 billion US$'s to 6,8 billion US$'s with increase of 92,2 % compare to 2003.

Parallel to increase of total production, an increase has been watched in export by 42,2 % in compare to 2003 which was amounted from 2,833,0 million US$'s to 4,029 million US$'s. It can be said that Turkish Electronic Industry produce mainly toward to exports. In 2004 74,0 % (3,138,0 million US$'s) of the total export belongs to European Union Countries. England ranked top in these countries to which exports were made 705,9 million US$'s followed by Germany 690,4 million US$'s, Italy with 302,0 million US$'s Spain, 268,9 million US$'s, Holland with 266,9 million US$'s.

When we look at the import of Electronic Industry we observe an increase of 47,9% compared to 2003. (from 6,0 billion US$'s to 8,9 billion US$'s) In 2004, 77,9% (4,212,0 million US$'s) of total import belongs to European Union Countries. Germany with 973,9 million US$'s, France with 459,8 million US$'s, England 286,7 million US$'s, Italy with 409,6 million US$'s.

Since the most of components such as IC used at electronic industry imported from other countries, import value is always higher than the export value. In early 1980's, a microelectronic fab called TESTAS in Ankara for production of integrated circuits had been planned to establish and for this purpose, a know-how for 7mm linear bipolar technology bought from EXAR Inc. in USA. In 1985, construction of fab had been completed but fab didn't start to pruduction and then project were cancelled. But, there are some design houses founded in Turkey.

The unique integrated circuit production facility in Turkey is present as a part of National Research Institute of Electronics and Cryptology (UEKAE) which is one of the institutes of Scientific and Technical Research Council (TUBITAK) of Turkey. Main areas of interest in UEKAE have been defined to be Information Security, Microelectronics and Optoelectronics considering the needs of information age and the importance of information.

Whether used for national or commercial purposes, secure storage or transmission of information which names the coming age is only possible by effective security measures.

As a result of this and the projects for the development of all sorts of communication security equipment using only national resources at UEKAE, Data Encryption Equipment, Secure Voice Equipment and special use communications equipment which could so far be only imported have been developed and offered to the use of the needing authorities.

The mission of UEKAE is "to conduct activities in the fields of Information Security and Advanced Electronics Technologies in order to contribute to technological independence of Turkey". In order to realize the vision of "To become an internationally recognized centre for science and solutions in the above mentioned fields with a solid infrastructure and highly qualified staff" determined with our mission in mind and to be able to develop technologies needed by the nation, basic and applied research are carried out in accredited test platforms and laboratories and technical support is provided upon demand.

UEKAE produces projects in partnership with public and private sector organizations and ensures participation of universities in these projects. In addition, the Institute responds to the national needs in the fields of information security and advanced electronics technologies, while contributing to the Post Graduate/Doctoral studies as well as universal science. Young researchers always have the chance to exploit their creativity fully by forming project groups under the leadership of a project manager.

UEKAE administration and employees believe fully in total quality and work constantly with this objective in mind for turning total quality management into reality within the Institute. Every UEKAE employee receives a total quality management training in order to achieve this. As a result of these efforts, the Institute has been awarded the ISO 9001-2000. With its understanding to find solutions to not only Government's but also private sector's information security needs, thus expanding its customer profile, the UEKAE aims to provide employment opportunities to young researchers of the nation where they can participate in proprietary research and development activities.

Microelectronics Research Laboratory (YITAL) of UEKAE was established in 1983. In YITAL founded within the research and development activities in the field of microelectronics technologies, a totally proprietary process has been developed and a silicon integrated circuit production line has been established.

An Overview of YITAL's Research Activities

At the beginning, the laboratory was equipped for the junction isolated 7µm bipolar IC technology, which was the established technology of early 1980's. With the NATO Science for Stability Programme codenamed TU-MICROFAB, the existing infrastructure and capability of YITAL was upgraded from standard bipolar technology to the CMOS technology in the period of 1988 and 1992. In the frame of this project, single metal, single poly 3-micron CMOS IC production process recipe was developed. One of the most important activities in this project was the procurement of second-hand ion implanter and GCA stepper because of the budgetary reasons. These equipments were installed and taken into operation by YITAL's researchers. Another important activity was upgrading the diffusion furnaces for wafer size from 3 inches to 4 inches.

1,5µm CMOS-ASIC Fabrication Technology Project, namely the TU MICROFAB-II project was included into the NATO-Science for Stability Phase III Programme and was started in 1994 and finalized in 1997. The main objective of this project was to design, simulate and optimize the steps of a twin-well, double metal, double poly-silicon 1.5 micron CMOS IC. The most important equipment added to the equipment park for this technology was PVD for metal deposition and PECVD for intermetal oxide. Turkish Electronics Industry contributed to the project giving financial support in providing some auxiliary equipment such as photoresist spinner and developer. At the end of the project, main objective was fulfilled successfully by gaining 1.5-micron CMOS technology.

TUBITAK-University Microelectronics Co-operation project was carried out by the cooperation of YITAL and five Turkish universities. In this project, it was aimed to motivate the students to microelectronics technology and IC design, and to encourage the university staff in courses on microelectronics. In the framework of this project, 45 chips were manufactured and given to designers for their evaluation.

0.7μm SOI CMOS process recipe was developed in 1999. In order to implement fully depleted SOI MOS transistors having low contact resistance between metal and silicon regions, recessed channel structure was realized.

In 2000, within an inner project of the Institute, laboratory was renewed to improve class-100 clean room conditions and was equipped for 0.7μm CMOS technology by adding a second g-line GCA stepper, a pattern generator for mask making on 5-inch chrome plates and a new cluster system for etching. Cluster system is equipped with two ICP (Inductively Coupling Plasma) modules for metal and poly-silicon etching, one RIE (Reactive Ion Etching) module for oxide and ICT (Down Stream Plasma) for photoresist strip off after metal etching. It has also a wafer dealer and load lock modules. ICP modules have capability of etching down to 0.35-micron future size.

YITAL's current infrastuctural capacity is given in Tables 10.1–10.2) for design and production. The facilities provide as self-contained production capability from full custom design to packaged integrated circuits; including all the necessary steps needed for mask making, wafer processing and testing.

YITAL design group is one of nine participants of EU-FP6-STREP project called as "Side Channel Analysis Resistant Design Flow-SCARD, with the contract number: IST-2002-507270. Collecting information in order to retrieve secret data from a cryptographic hardware using the signals that leaks through one of its side-channels such as supply current or electromagnetic emanation is called side-channel analysis (SCA). The goal of the project is to develop design technics, which prevent information leakage in smartcards via side channels. The role of YITAL design group in this project is to develop a secure SRAM structure whose power consumption is independent from the data values read from or written into the memory and the address values accessed.

Overall Process Sequence

In the production of integrated circuits, four main processes are applied on silicon wafers. These are: (1) photolithography, (2) ion implantation, (3) deposition of layers, (4) etching. In the optimization of CMOS process steps, we have only benefited from open literature. For evaluating the performance, reliability and reproducibility of the process sequences and to develop the design rules for 1.5 μm CMOS technology, many experiments were carried out.

In this section 1.5 mm double metal CMOS process sequence optimized in YITAL is described. In a CMOS process, there are 5 main steps.

WELL Formation

Twin-well structure enabled the independent control of the threshold voltages of both NMOS and PMOS transistors from the resistivity and polarity of substrate.

In order to decrease the number of lithography steps in the overall process sequence, self-aligned technique, which is based on the local oxidation of silicon, was applied in the formation of the n-well regions. First, silicon wafers were oxidized to grow 45nm pad oxide and then silicon nitride, 140nm in thickness, was deposited by LPCVD technique. This silicon nitride layer, after having been patterned by p-well mask, was etched in the parallel plate plasma system with CF_4 and O_2 gases. This was followed by a high energy, low dose implant of boron to form the source for the p-well diffusion step. Following photoresist layer removal, silicon wafers were oxidized at 1000 °C in pyrogenic system (H_2 and O_2) to grow 500nm masking oxide at the surface of the p-well regions, silicon nitride layer was stripped dipping in H_3PO_4 wet etchant and phosphorous ions ($^{31}P^+$) were implanted to form the source for n-well regions. Drive-in cycle was performed at 1125 °C in N_2 for 30 h in order to realize the n- and p-well diffusions. Since the yield of a process is inversely proportional to the number of the crystal defects generated during the fabrication sequence, a low temperature (1125 °C) process was applied in order to avoid slip dislocation and oxidation induced stacking fault formation during the drive-in cycle. Sheet resistance of n- and p-well regions at the end of the drive-in cycle were 1800 W/o and 450 W/o respectively.

Isolation

Conventional LOCOS process was applied for isolation purposes in 1.5μm CMOS process. The stack, consisting of a pad oxide and silicon nitride was formed with the thickness of the films were 45nm and 140nm respectively. Following the formation of the pad oxide and silicon nitride layer, nitride layer was patterned with active area mask and etched in CF_4+O_2 plasma. This was followed by the second patterning step with p+ field mask, in order to implant the field regions with boron at 25 keV to increase the parasitic threshold voltage in these regions of the integrated circuits. Following the field implant, a 550nm field oxide was grown at 1000 °C in pyrogenic system (H_2, O_2), and then nitride was stripped in phosphoric acid (H_3PO_4) at 150 °C.

Transistors in the same well region are isolated electrically from each other by the combination of a thick field oxide and a high concentration surface layer under the field oxide. The structure results in a high threshold voltage for the parasitic transistors. This threshold voltage should be higher than the maximum voltage, which is applicable to the polysilicon gates of the parasitic field devices inevitably formed by the interconnection of devices. Hence, the field oxide thickness depends on the circuit design requirements. A thicker field oxide such as 800nm would be necessary to reduce parasitic capacitance and increase parasitic threshold voltage. But a thin field oxide (550nm) can be tolerated with the assumption that no longer polysilicon interconnects are used in a process with two metal layer. Decreasing the number and size of the oxidation induced crystal defects and keeping the bird's beak encroachment as short as possible was the main concept behind the selection of a thinner field oxide. It also helped to lower down the oxidation temperature. At the end of the LOCOS process, the length of bird's beak encroachment was measured to be 0.6µm (±0.1µm), which was acceptable for 1.5µm technology design rules (Figure 10.1).

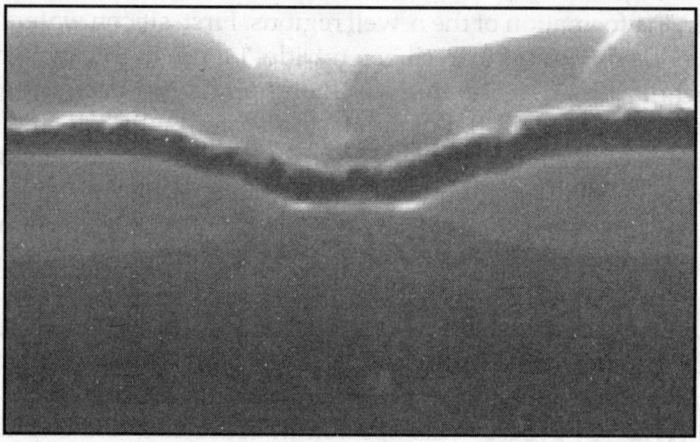

Figure 10.1: SEM Photograph of Bird's Beak Encroachment

MOS Devices

In order to perform threshold adjust implantation, a sacrificial oxide, 25nm in thickness, was thermally grown and low energy (25keV), low dose (1.55 e12cm^{-2}) boron implantation was performed to have threshold voltages of NMOS and PMOS transistors +0,6 V and -0.6 V respectively. The p-well and n-well implantation parameters and the drive-in step have been so arranged that adjusting the threshold voltages of two different types of transistors with a single implantation step was made possible. Threshold adjust implant dose for double poly process sequence was increased to 1.75e12 cm^{-2} to overcome the boron depletion at the channel region of MOS devices and keep the threshold voltages at ± 0.6V.

Following threshold adjust implantation, sacrificial oxide was removed from the surface and an RCA cleaning was applied on the silicon wafers before the growth of the gate oxide, 25nm thick, in dry ambient at 900 °C. Experiments were also carried out to grow the chlorinated gate oxide by adding TCA (thrichloroethane) as chlorine source to the oxidation ambient and the results, obtained by the simultaneous triangular voltage sweep (STVS) method, pointed out that the alkaline (mobile) ion concentration in gate oxide decreased to a value of 3.7E9 cm^{-2} from 2E10 cm^{-2}. Although the mobile ion concentration in chlorinated oxide was 10 times lower than that of in dry oxide, dry oxidation was chosen to grow the gate oxide because, not only alkaline ion concentration of dry oxidation was within acceptable limits but also dry oxidation needed a simpler process recipe than chlorinated one. Since gate oxide has a vital importance in IC fabrication, it requires precise thickness control. The thickness uniformity is about 3% both within a wafer and a run.

After the gate oxide growth, undoped polycrystalline silicon (polysilicon) was deposited by pyrolyzation of silane (SiH_4) at 600 °C in LPCVD furnace. Phosphorus doped oxide (PSG) was used as the phosphorus source for doping the polysilicon layer. In this application, PSG film was deposited by TEOS based LPCVD system at 660°C and phosphine (PH_3) was used as the dopant gas. The deposition of PSG on surface of the polysilicon layer was followed by the drive-in step, at 925°C in nitrogen ambient, which is needed for the diffusion of the dopants in PSG film into polysilicon in order to make the sheet resistance of polysilicon 20 ± 5 W/o. Doped polysilicon layer was etched in RIE system by the gate mask. Patterning and etching of polysilicon is a critical step with a ± 0.1μm tolerance at 1.5 μm features. The polysilicon was processed by an anisotropic plasma etch with high selectivity to both the resist mask and the underlying gate oxide. This polysilicon etch process, performed by RIE, should exhibit good uniformity to minimize the over etch time.

After patterning of polysilicon has been completed by the etch process, photoresist was removed and wafers were cleaned using RCA. Since gate oxide could also be somewhat etched during the polysilicon etch, a reoxidation cycle (post oxide-1) was performed to increase the thickness of gate oxide under the polysilicon electrode edges in order to maintain the integrity of the thin gate oxide. This post oxidation also results in annealing which cures some, but not all, crystal damages occurred during polysilicon etch in RIE system.

Source/Drain Region Formation

In 1.5μm CMOS processes, the gate lengths are short enough to create high electric field near the drain regions of transistors and lightly doped drain (LDD) structure was used to minimize the hot-carrier effects in NMOS devices. In 1.5μm CMOS technology having LDD structure, the drain regions of NMOS transistors were formed by two implants. First one, called the LDD implant, was self-aligned to the polysilicon gate to form the lightly doped section of drain at the edge near the channel and the implant dose was 4×10^{13} cm^{-3} of phosphorus. Photoresist layer patterned with LDD mask against to phosphorous ion implantation protected PMOS transistors. In order to form oxide spacers at the two sidewalls of polysilicon electrodes, undoped TEOS oxide having a thickness of 250nm was deposited in LPCVD system and it was etched in RIE system without any lithography step. Hence, spacer formation was purely dependent on the anisotropy of the etching process. During the oxide spacer etch step, screen oxide over the active area and field oxide were also etched. The etch cycle should be applied by taking into account the uniformity of the spacer oxide thickness. Over etching of the oxide on the active regions should be prevented for device isolation and junction leakage. Following the formation of the oxide spacer, arsenic ions were implanted at 120keV as the second implantation to form highly doped part of the drain regions of NMOS transistors. Dose of implantation was $6.10^{15}cm^{-2}$. The drain regions of PMOS transistors were created by a single implantation step where boron ions were implanted at 25 keV.

The aging tests were carried out to determine the hot-carrier reliability of the 1.5μm CMOS process. It was observed that the amount of the hot-carrier damage in the test transistors was relatively low compared to those reported in the literature, which shows that, LDD structure parameters (the length of oxide spacer, LDD implant dose and energy) were successfully determined to prevent the hot-carrier damage.

Metallization

BPSG film, 1.1μm thick, was deposited by TEOS based LPCVD system as an isolator between polysilicon and metal-1. To get better step coverage for undoped SiO_2, a densification cycle was realized at 925 °C in N_2 ambient. This stage also activates the n^+(S/D), p^+(S/D) and threshold adjust implants. Contact opening is another critical step at which a ±0.25μm etched feature specification has to be met. The contacts were etched through the silicon dioxide film using a two step etch process. The first step was an anisotropic etches in RIE system with high selectivity to the resist mask. The oxide having about 800nm thickness was etched by this step and a wet etching (BOE) was applied to etch down to the silicon surface. This two-stage process reduces the amount of crystal defects caused by RIE process and decreases the contact resistivity. Following the metal-1 deposition and patterning, a sintering-1 (alloy) step was performed at 425 °C in N_2 ambient. Intermetal oxide (IMOX) was deposited by PECVD system. A four step process including TEOS deposition-1 (500nm), sputter etching by argon plasma, TEOS deposition-2 and a reactive chemical etching by CF_4 and O_2 plasma was realized to obtain better planarization of IMOX. Final thickness of IMOX was 1,1μm. Via etching was performed by RIE system and Metal-2 was deposited and etched. Following the sintering-2 step having the same conditions with sintering-1, a passivation layer was deposited by pyrolizing SiH_4 and O_2 at 400 °C in APCVD (Atmospheric Pressure Chemical Vapor Deposition) system. Pattering of passivation layer completed the production run.

Table 10.1: Manufacturing Equipment

Equipment Name	*Quantity*
GCA Wafer Stepper	2
Dry Etchers	
ICP	2
RIE	3
Plasma	2
PVD	1
PECVD	1
Ion Implanter	1
LPCVD	3
Furnaces	9
Mask Generator	1
SEM (JEOL)	1.

Contd...

Table 10.1–Contd...

Equipment Name	*Quantity*
Wafer test equipment as follows:	
Automatic wafer prober	1
Manual wafer prober	1
Semiconductor Parameter Analyzer	2
ASIC tester	1

Table 10.2: YITAL has the following Infrastructure for Design

EUROPRACTICE Membership
CHIP DESIGN SOFTWARE
CADENCE
Schematic Design
Layout Design
Verification (DRC, ERC, Extract, LVS)
VERILOG-Digital Simulation
SYNOPSYS (Circuit Synthesis and Simulation Software)
HSPICE-Analog Simulation
NANOSIM-Analog and Digital Simulation
CHIP DESIGN HARDWARE
Workstations
Personal Computers with operating systems of MS-Windows and Linux

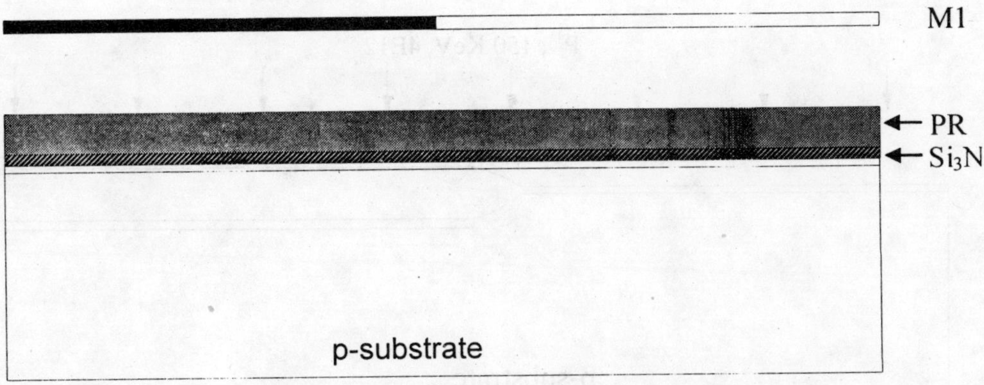

1. Pad Oxide Growth, Nitride Deposition and P-Well Patterning

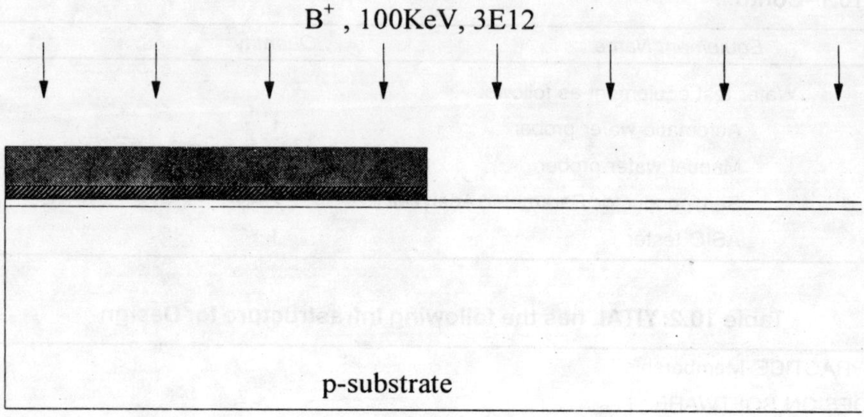

2. P-Well Implant

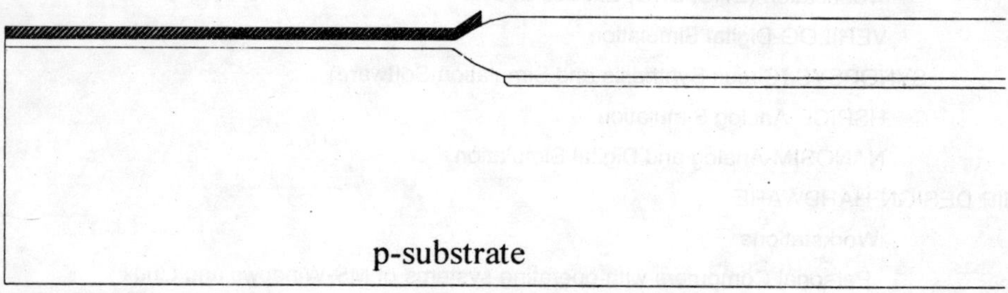

3. Oxidation of P-Well Regions

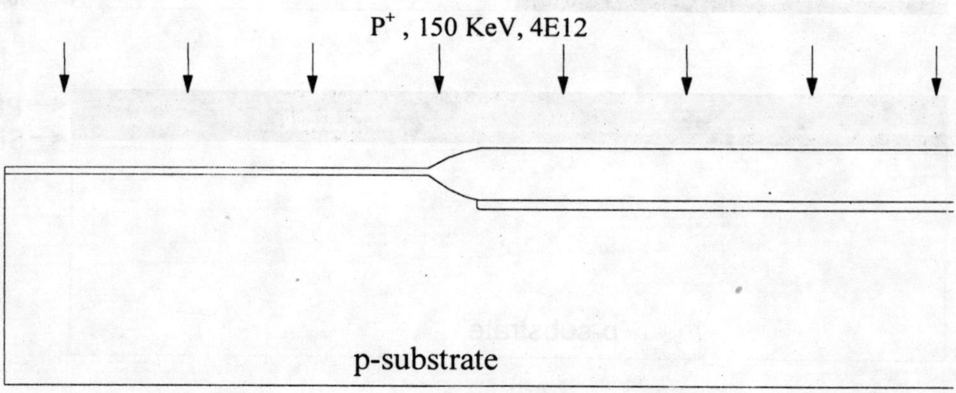

4. N-Well Implant

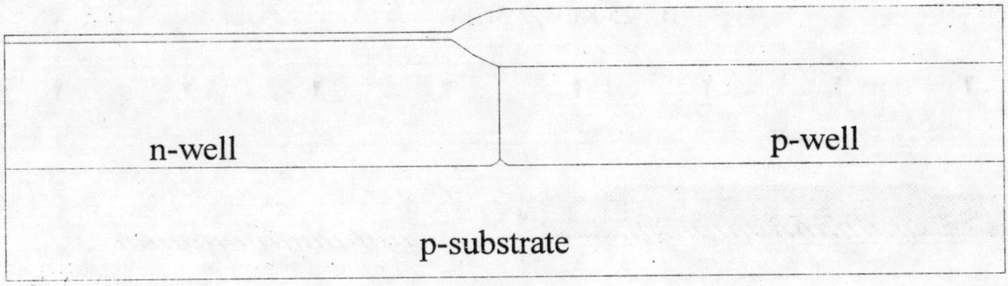

5. Well Drive-in

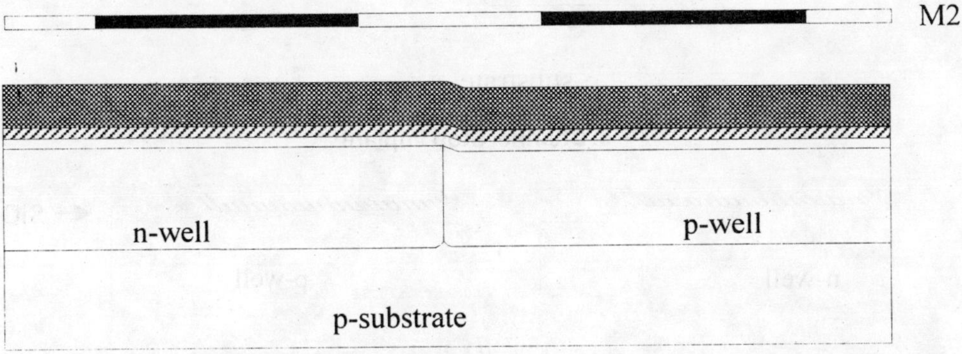

6. Oxide Etch, Second Nitride Deposition and Active Area Patterning

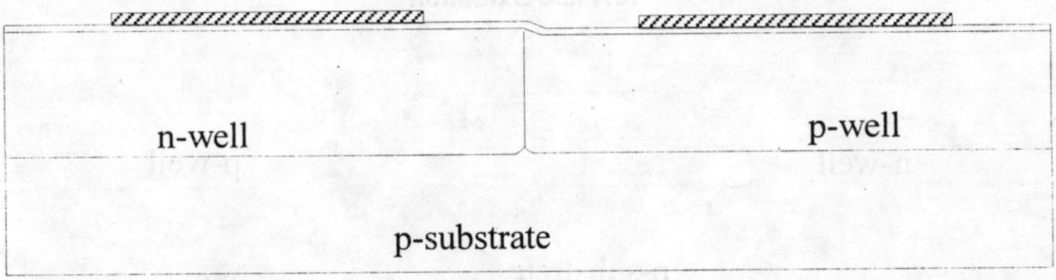

7. Nitride Etch

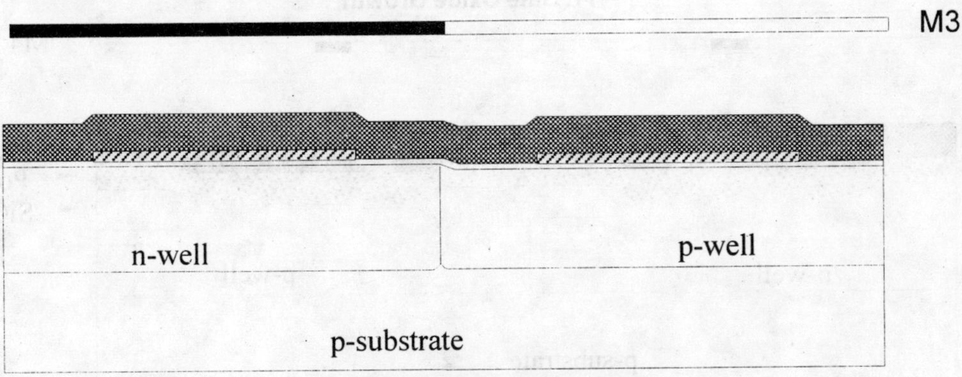

8. P+ Field Regions Paterning

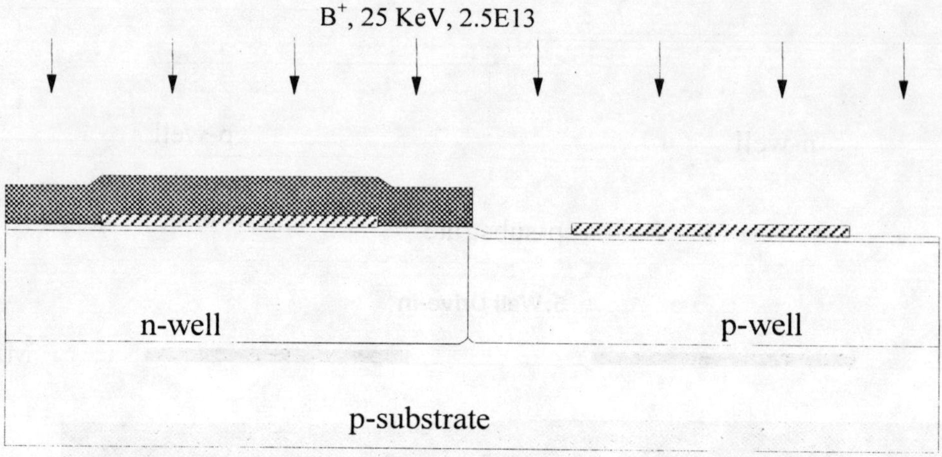

9. Channel Stop Implant

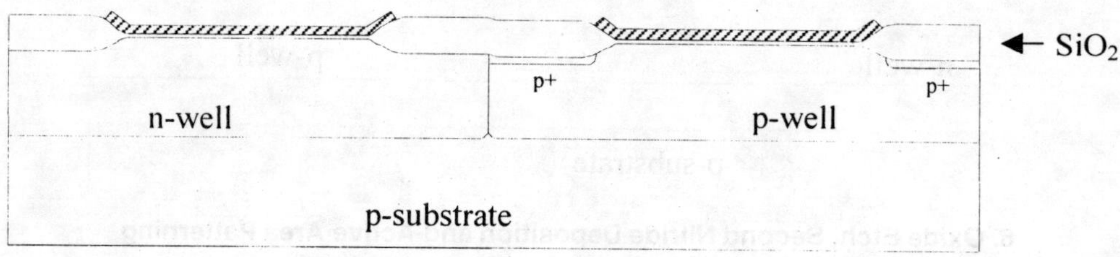

10. Field Oxidation

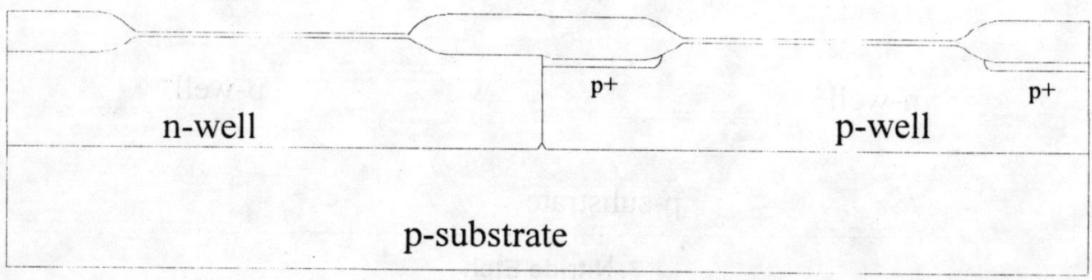

11. Gate Oxide Growth

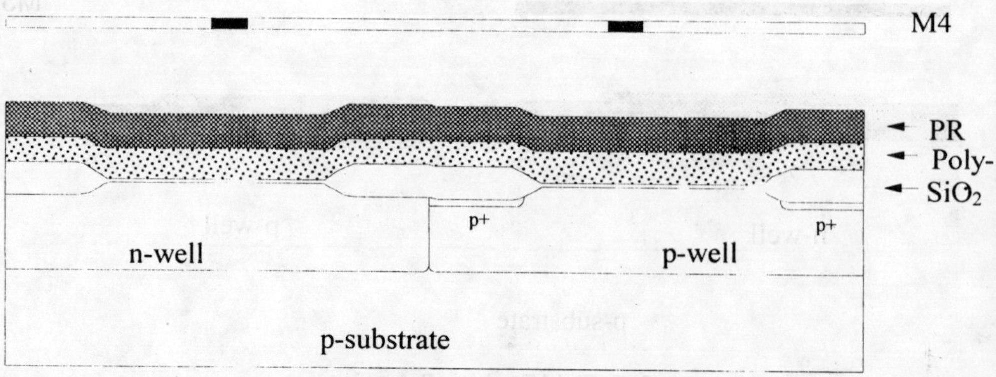

12. Poly-Si Deposition, Doping and Patterning

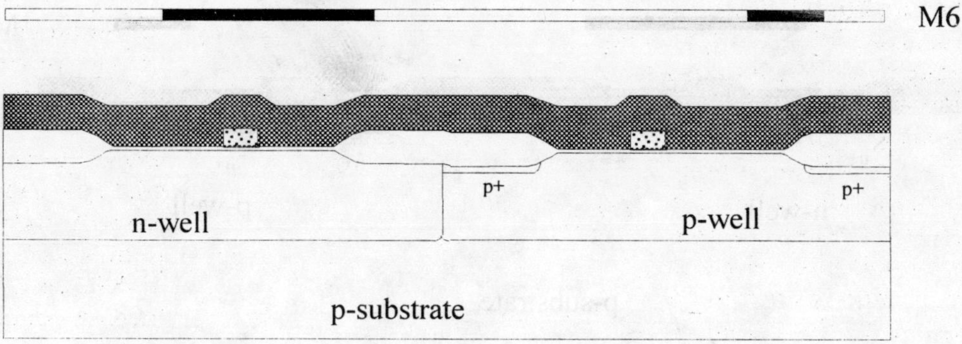

13. LDD Regions Paterning

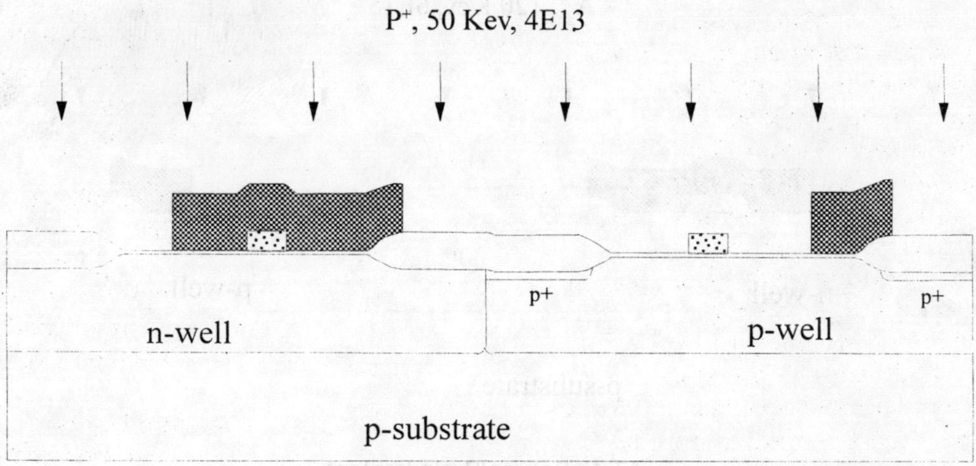

14. N-Implantation for LDD

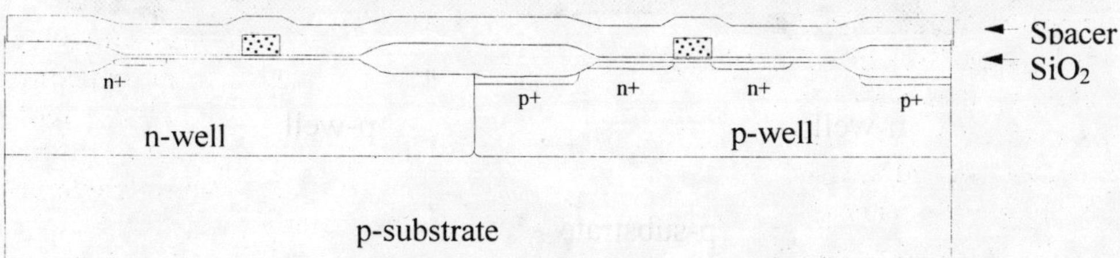

15. Spacer-Oxide Deposition

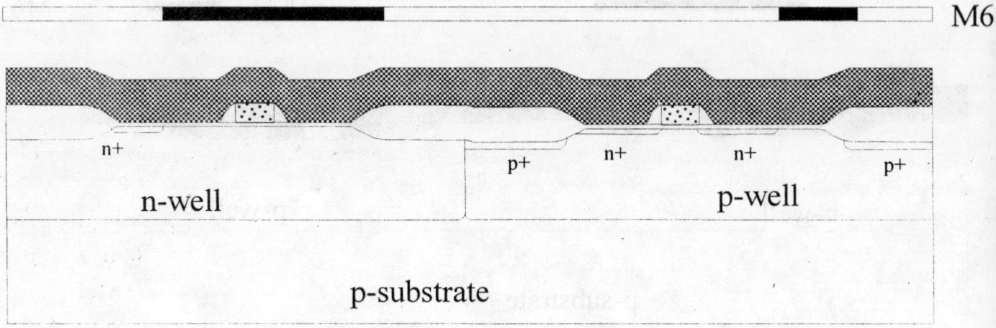

16. N⁺ Source/Drain Regions Paterning

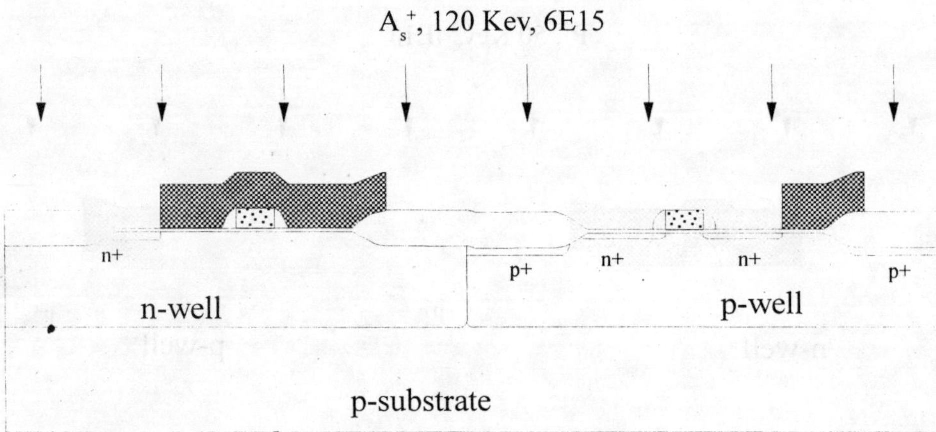

17. N⁺ Source/Drain Implant

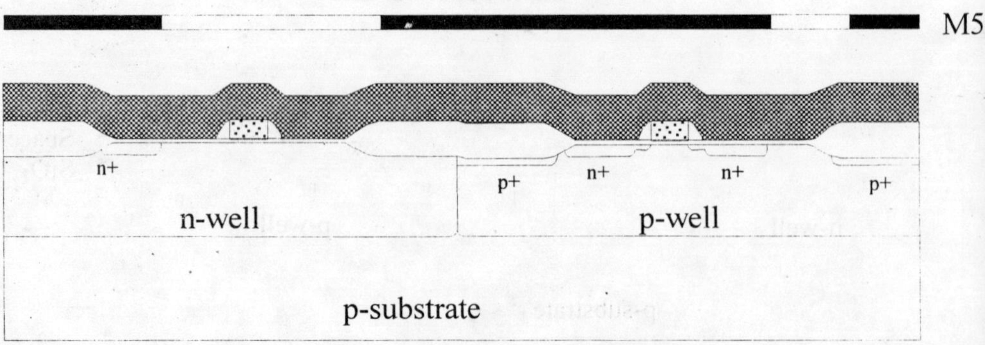

18. P⁺ Source/Drain Regions Paterning

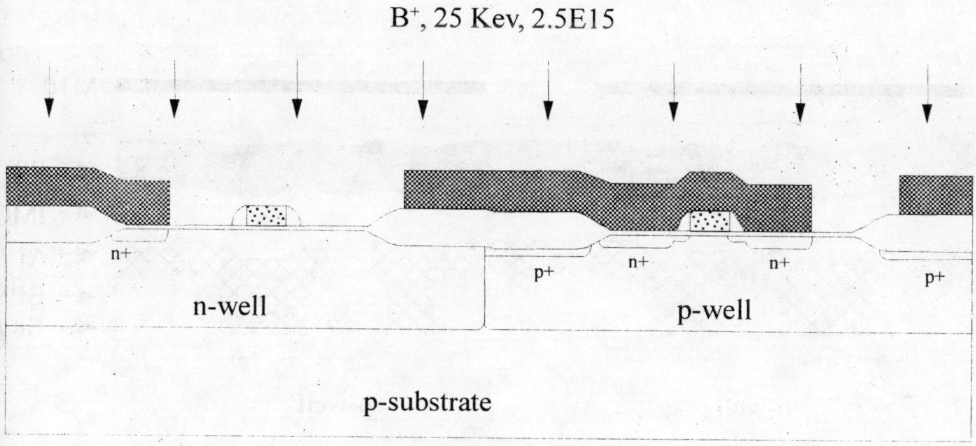

19. P+ Source/Drain Implant

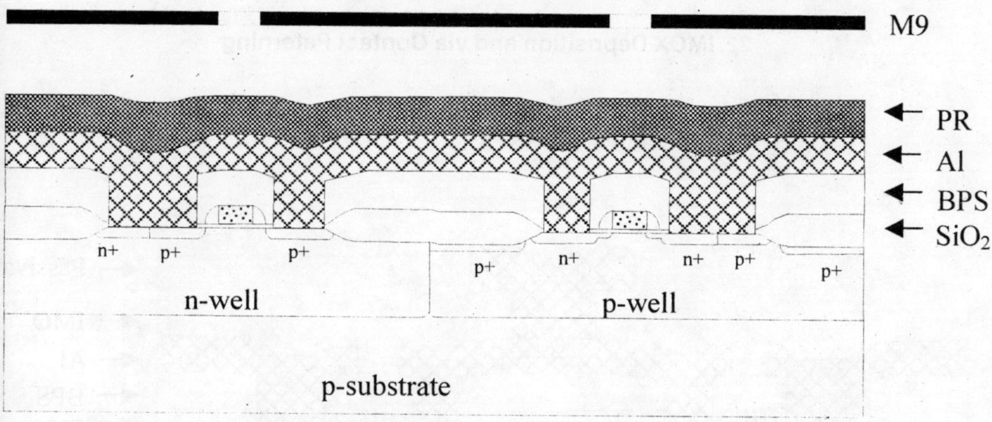

20. BPSG Deposition and Contact Paterning

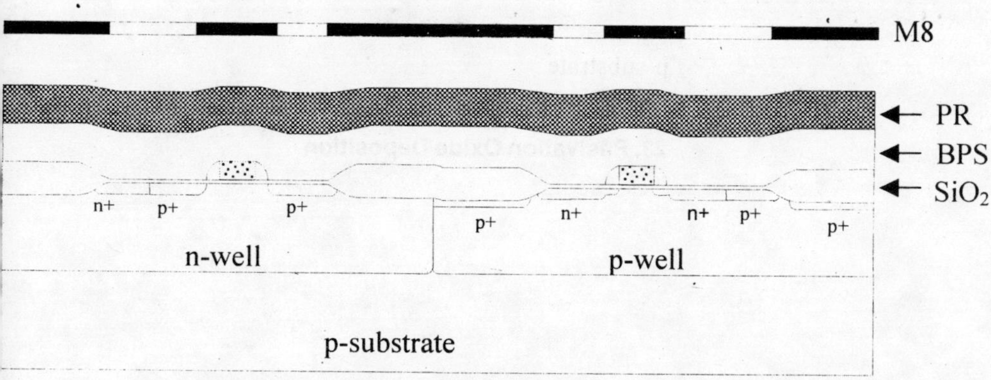

21. Aluminium (Metal 1) Deposition and Paterning

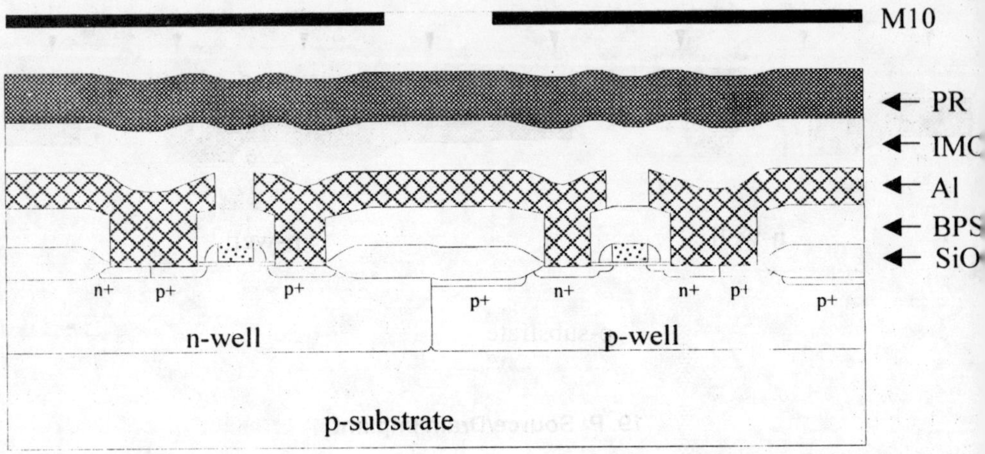

22. IMOX Deposition and via Contact Paterning

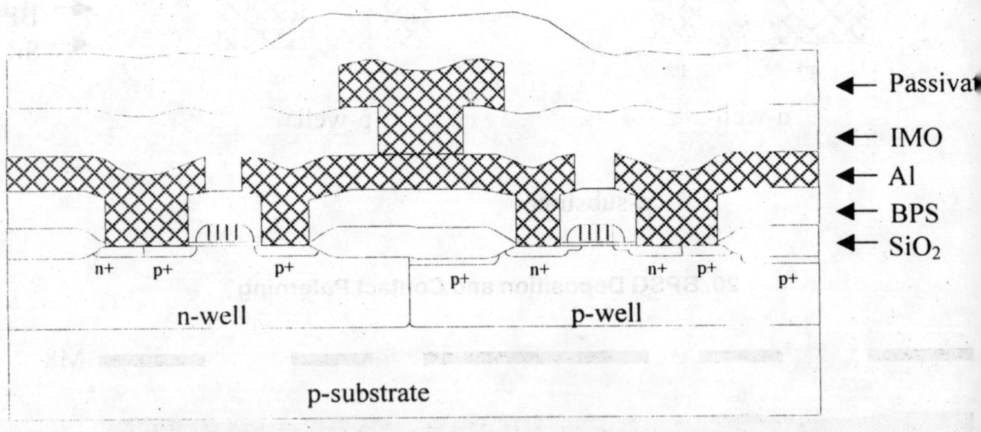

23. Pasivation Oxide Deposition

11

Development and Applications of Microelectronics in Nigeria

Ashiru S. Daura
National Information Technology Development Agency, Federal Ministry of Science and Technology, Nigeria
E-mail: asdaura@nitda.gov.ng

Introduction

Research and Development in Microelectronics in Nigeria is not very significant. This is due a number of factors affecting the Universities and Research Institutes, and the Industries. However, the recent interest shown by government, through its reforms, gives some hope and encouragement.

The Federal Government has declared three areas of high technological importance as its priorities. These areas are: Biotechnology, ICT and Space Technology. To support this resolve the Government has promulgated a national policy for each as well as established an agency each to see to the implementation of the respective policies.

Current Situation

There are a number of problems that retard R & D efforts. These include lack of sufficient funding, lack of private sector participation, brain drain and the culture and habit of importation of finished products. Nevertheless, this section highlights some areas of activity.

In more than one hundred Universities and Polytechnics, Electronics Engineering and Physics are taught, and basic researches are carried out by lecturers and students.

However, most of these researches are neither published nor developed due to some of the problems mentioned above.

In addition to the Universities and Polytechnics, government has established a research institute that specializes in Microelectronics R & D. This is the CENTRE FOR ADAPTATION OF TECHNOLOGY (CAT), which is a unit under the National Agency for Science and Engineering Infrastructure (NASENI). The Agency is under the Federal Ministry of Science and Technology. The mandate of the Centre is to develop and produce electronic devices and assemblies, computer technologies and their production systems, then transfer these technologies to private sector satellite industries. The Center's current projects include:

- Development and Production of Printed Circuit Boards
- Research and Development of Electric and Solar Powered Inverters
- Design and Development of Equipment for Generation, Transmission and Reception of Signals in the Ultra High Frequency (uhf) Band
- Systems Programming for Design and Development of Electronic Embedded Systems

It is hoped that the Center will, in the near future, be involved in more and rigorous state-of-the-art Microelectronics R & D.

A major area in the current application of Microelectronics is the Global System for Mobile communications (GSM) deployment. From year 2000 to date the number of GSM lines has risen from nothing to over four million. This has brought up a new breed of Engineers and Technicians specializing on the miniature circuits of the handsets. The recent move by some companies to invest in local manufacturing of handsets and government directives in that direction through the regulatory National Communications Commission (NCC) would no doubt trigger a revolution that will encourage the development of microelectronics to boost Nigerian local content in the ICT industry.

The Shape of the Future

As seen in the preceding paragraphs, R & D activities in Nigeria are far below expectation. However, the recent interest shown by government, through its reforms, gives some hope and encouragement. The government has identified three key areas –Biotechnology, ICT and Space research- as its priorities. Definitely, a faithful development of these areas has to be accompanied with a tremendous amount of activity in microelectronics development and applications. For this reason a few paragraphs will be devoted to reviewing two of these areas.

Information and Communications Technology

The Federal Executive Council approved a national IT policy in March 2001 and the implementation started in April with the establishment of the National Information Technology Development Agency (NITDA), charged with the implementation responsibility. The policy recognized the private sector as the driving engine of the IT sector. NITDA is to enter into strategic alliance, collaboration and joint venture with the private sector for the actualisation of the IT vision, which is to make Nigeria an IT capable country using IT as an engine for sustainable development and global competitiveness. It is also to be used for education, job creation, wealth creation, poverty eradication and global competitiveness.

Emphasis is to be laid on National Information Infrastructure Backbone (NIIB) as well as the Human Resources Development. In addition, Information Technology Parks are to be developed in Abuja and in each of the six geo-political zones.

The specific objectives of NITDA include the following:

- To ensure that Information Technology resources are readily available to promote efficient national development.
- To create IT awareness and ensure universal access in order to promote IT diffusion in all sectors of our national life.
- To ensure simple, moral, accountable, responsive and transparent governance
- To eliminate waste and ensure that governance and businesses are done in the global standard of using ICT for easier, faster and cheaper delivery of services.
- Transformation of Nigerian economy from resource based to knowledge based economy
- To encourage local production and manufacture of IT components in a competitive manner in order to generate foreign earnings.

NITDA has embarked on R&D activities to achieve the objectives of the National IT Policy. The major achievement includes the development of Mobile Internet Units (MIUs). The MIUs are luxurious buses, equipped with structured cabled local area network (LAN) of computers (servers and workstations), printers, and multi-media facilities. A transportable Very Small Aperture Terminal (VSAT) equipment with a very small dish mounted on the roof of the bus provides Internet access. One unique aspect of the MIU is that it is a completely made in Nigeria product through the initiative of NITDA. The Mercedes Benz buses used were assembled in Nigeria; the restructuring of the buses into mobile computer laboratory was done by Scientific Equipment Development Institute (SEDI) which is another Agency of the Federal Ministry of Science and Technology, while the structured cabling and installation of other IT facilities were done by NITDA.

Also the Agency through its R&D Unit had developed a Nigerian Computer keyboard to facilitate local content in the area of software development. The initial plan is developed to accommodate three major Nigerian Languages.*i.e.* Hausa, Igbo and Yoruba.

One very significant move towards R & D and capacity building in this area is the approval for the establishment of the first Technology Park (Abuja Technology Village – ATV) in the capital. Its mission is to provide world-class infrastructure, civic amenities and knowledge-driven environment to facilitate technological advancement, job creation, export diversification and hasten poverty reduction in Nigeria.

A number of state governments have also made IT policies and developed various levels of infrastructure in their respective states.

Space Technology

Recently, a National Space policy was developed, with a vision of making Nigeria build indigenous competence in designing, developing, and building appropriate hardware and software in space technology as an essential tool for its socio-economic development and enhancement of the quality of life of its people. To further this vision, the National Space Research and Development Agency was established.

The Agency has made a laudable landmark by launching the first Nigerian satellite, an earth observation micro satellite, NigeriaSat 1, into orbit in September 2003. It is also embarking on the next generations of satellites: a communication satellite to be called NigcomSat-1 and a high resolution African Resources Management Constellation (ARMC) satellite, NigeriaSat-2. NigcomSat-1 is to be launched in 2006.

NASRDA is also doing well in the area of remote sensing and astronomy through its various centres. Its R&D centres at a number of the nation's Universities are carrying out various R & D activities in space science and technology and their applications.

Reforms

All these developments are not sustainable without strong R & D, particularly in microelectronics, which has tremendous and essential applications in the areas mentioned, and which is necessary for self reliance and development, with a strong Public-Private sector Partnership (PPP). In realization of this, a lot of efforts are being made, the latest being the reform of the Science, Technology and Innovation systems of the country, launched in late July, in collaboration with UNESCO.

Elements of the reform programme include:

- International Review of the functions of Nigeria's National System of Innovation
- Development of a financing system for S&T
- In-depth review of selected S&T institutions
- Human Resources Development Program for S&T Policy and the Management of Technical Change
- A program on information management
- Science, technology and innovation in Nigeria's private sector
- A program to strengthen Nigeria's infrastructure in the area of standards and metrology
- A review of Nigeria's system for the protection of intellectual property
- A review of Nigeria's participation in regional and international scientific activities

Conclusion

Research and Development in Nigeria, particularly in microelectronics, is very much at a low level. However, current developments give hope of a boost in the near future.

12

Development of Microelectronics and Microsystems in Pakistan

Muhammad Asif

Assistant Professor

COMSATS Institute of Information Technology, Islamabad, Pakistan

E-mail: massif@comsat.edu.pk

Introduction

Pakistan, a land of many splendors and opportunities, repository of a unique blend of history and culture of the East and West, the cradle of one of the oldest civilizations *i.e.*, the Indus valley, is the ninth most populous country of the world with 125 million tough, conscientious, hard-working people wishing and striving hard to enter into the 21st century as equal partners in the community of developed nations.

Location

Latitude: between 23 and 37 degrees north.

Longitude: 61 and 76 degrees east.

Iran and Afghanistan in the west and Central Asian republics and China in the north and sharing its more than 1000 Km border with India- Pakistan can rightly boast of having a significant location advantage with a vast, tapped and un-tapped market of 200 million people. The affluent Gulf states are just across the Arabian Sea in the south that provides an additional vista of a high mass consumption market.

We have:

(a) Political commitment,

(b) will for Indigenization and self-reliance,

(c) An organized program for Human resource development, and

(d) Internationalization.

Pakistan, spread over a land mass of 796,095 square kilometers has vast, relatively cheap land for setting-up industries. The average office occupation cost in the prime business districts of the big cities at less than US $10 per sq. ft. per year is much lower than other countries of the region.

Development of Microelectronics and Microsystems in Pakistan

Educational and Teachings

There are approximately 50 universities which are providing education in microelectronics at undergraduate level, and some 15 universities have been involved in teaching at postgraduate level up to M.S. and Ph.D. level. The main concern of these universities is research in the areas of Semiconductor and Microelectronics.

Some of the important research activities are given as under:

Quaid-i-Azam University, Islamabad

The University involved in research on 3-5 compound in semiconductors and has produced several Ph.D. graduates and published important research publications. The QU has well equipped laboratories having facilities of characterization technologies and undertake research in basic physics of 3-5 compound semiconductors.

University of the Punjab, Lahore

It mainly involved in the research in the areas of characterization of semiconductors and other materials.

Mehran University of Engineering and Technology, Jamshoro

It is established at Jamshoro, Sindh to produce much required highly qualified manpower at various levels of policy, management and administration of promotion and development of Science and Technology Enterprise in private and public sectors. Short training, M.S., M. Phil, and Ph.D. programs in science and technology policy and development are the primary targets. It is carrying tasks of review and conducting research on important current and futuristic issue pertaining to science and technology policy and development and of interest to the academia, institutions, industry, business and the Government to achieve rapid socio-economic development in all sectors of development of the country. In particular, it aims to ensure successful co-operation between government, industry and universities to achieve effective co-ordination among Education, Science and Technology and Industry. It envisages further to establish Scientific and Technological information system and networking for statistical analysis, dissemination, research and decision making.

National University of Engineering and Technology (NUST), Islamabad

NUST offers bachelor program in engineering disciplines. Presently the university offers undergraduate programs in Aerospace (Aviation Mechanical) Engineering, Avionics (Aviation Electronics) Engineering, Computer Engineering, Software Engineering, Information and Communication Systems Engineering, Telecommunication Engineering, Electronics Engineering, Mechanical Engineering and Mechatronics Engineering. The university also offers MS Engineering programs in various engineering disciplines, including Aerospace Engineering, Avionics Engineering, Electronics Communications, Control Systems, Software Engineering, Communication and Signal Processing, Solid State Electronics, Microwave Engineering, Telecommunication and Information Security.

Sir Syed University of Engineering and Technology, Karachi

It has very good undergraduate programs in Bio-Medical Engineering, Electronic Engineering, Computer Engineering and post graduate teaching and research programs in Computer Networking, Software Engineering, Telecommunications, Industrial Electronics, Mobile and Wireless Network and Management.

Islamia University, Bhawalpur

It is actively involved in research areas of characterization of semiconductor, particularly in the area of Temperature- dependent Admittance Spectroscopy (TAS).

Polytechnic and Technical Education Institutes

In addition there is a huge demand of technical and skilled manpower in the country. In order to meet this requirement Govt and the private organizations have established hundreds of Polytechnic and Vocational institutes. Most of them are providing basic electronic education and training in electronics for the commercial and industrial needs.

Research and Development Organizations

There are different R & D organizations carrying out research and development on microelectronics and microelectronics devices. Major activities are being undertaken at the following organizations:

National Institute of Electronics (NIE), Islamabad

NIE is the organization under the ministry of science and technology conducting research and development in different areas of electronics. It has two main divisions

1. Component Division
2. System and Circuit Division

Component division started its programs in PCB designing and manufacturing and is now capable of producing state of the art PCBs on small scale. This division has also established state of the art CAD facilities for semiconductor chip design. A process research lab is also established for the training to the university graduates in different level of silicon processing.

The system and circuit division carries out research and development work in the area of automation and control engineering, embedded software and hardware, customize designing of communication equipment.

Pakistan Council for Renewable Energy Technologies

The primary concern of this institute is purification and development of silicon wafers. The institute has the facilities for purification of silicon to metal grade and semiconductor grade silicon. PCRET is also capable to produce silicon single crystal wafers in 10 cm diameter.

Owing to the activity in non-conventional power generation the institute has undertaken manufacturing of solar cell and solar panels. The institute has two main divisions.

1. **Material Processing and Cell Fabrication Department** is subdivided into Crystal Growth Group, Wafering Group and Lamination Group. The Crystal Growth Group used to grow single crystal of silicon (10 cm diameter and up to 1 m long). Wafering Group has the facilities to cut the silicon crystal into 500 micron thick wafers for subsequent processing into solar cells. Finally the Lamination Group is responsible for linking the solar cells together and then encapsulated between layers of glass and polymers.
2. **Cell Processing Division** is sub grouped into Solar Cell Fabrication Group, Thin Film Group, and Cadmium Telluride Group. The main functions of the Solar cell fabrication Group are Chemical etching of the wafer surface, screen printing of aluminum paste on one side of wafer, edge grinding and cleaning. The Thin Film Group is responsible for routine deposition of antireflection thin film, and also research on related issues such as alternative antireflection materials, transparent conduction coatings, and Metal-Insulator- Semiconductor (MIS) cells. The third group is interested in the back contact formation using zinc-telluride etc.

Career Telephone Industries (CTI), Pakistan

CTI is producing VHF/UHF and other communication equipment for telecom sector. The industry has large production facility also it has discrete device fabrication and packaging facility. The R&D section of CTI developed many Telecommunication products like VHF/UHF sets, Digital radio systems, MUXs and other communication products.

Telephone Industries of Pakistan (TIP)

TIP has been major industry involved in the production of communication equipment (mainly telephone sets) for Telephone and Telegraph (T&T) department. It has been meeting almost 90% of the demands of the country. Now due to privatization process in the country T&T has been converted into Pakistan Telecommunication Co. Ltd (PTCL). The other private companies have come in and TIP's market share is about 40%.

Central Telecommunication Research Labs.(CTRL), Islamabad

CTRL has been conducting research and development in the areas of digital and analog communication systems and components. The major areas of research have been VHF/UHF/Microwave, Passive components and materials, semiconductors and thin film/thick film circuits. The Lab has conducted valuable research on ceramic materials, semiconductors and VHF/UHF/Modems. The activity is minimal now a days due change in structure from T&T to PTCL.

National Radio Telecommunication Corporation (NRTC), Pakistan

NRTC is involved mainly in the production and research of Telecommunication equipment. The corporation has been one of major supplier to PTCL in the telecom sector. The market share of the industry in telecom sector has been about 30% on the average.

Private Sector

The main activities of private sector are in the telecom sector and consumer electronics. There is boom generally in communication and particularly mobile communication. There is record growth in mobile communication and IT sector within last 2-3 years.

In microelectronics particularly there is minimal activity. However there is lot of potential to participate and take part in development of microelectronics sector in the country. There is huge potential market for this sector in Pakistan. In telecom and consumer electronics sector lot of investment has been made by local and foreign investors. It is matter of educating and encouraging private sector and foreign experts/ companies/investors to come forward and mutually benefit from development in microelectronic sector.

Microelectronic Activities at COMSATS Institute of Information Technology, Islamabad

The Commission on Science and Technology for Sustainable Development in the South (COMSATS) is an international organization. It aims to reduce the ever-growing gap between the developed and developing world through useful applications of science and technology.

The Third World Academy of Sciences (TWAS) initiated the proposal for the formation of COMSATS under the leadership of Nobel Laureate, Prof. Abdus Salam. The foundation-conference of COMSATS was held at Islamabad on 4th & 5th October 1994. Representatives from thirty-six countries attended. The participants included twenty-two Ministers, members of the diplomatic community of Islamabad and representatives of international organizations, like UNESCO, UNIDO, UNEP and the World Bank. Along with other modalities the conference decided that the Headquarters/Secretariat of the Commission would be based permanently at Islamabad.

COMSATS has selected thirteen Centers of Excellence in its member countries to develop a Network of International Science and Technology Centers for Sustainable Development in the South. These centers have been selected as nodes and are being used to provide a leading role in their respective areas of specialization.

Within Pakistan major activities of COMSATS are taking place through its subsidiaries; COMSATS Internet Services (CIS) and COMSATS Institute of Information Technology (CIIT). CIS has set up and is operating modern communication systems in most cities throughout the country and CIIT has been established for training professional human resource. CIIT is imparting education to over 10000 students nationwide. With its headquarters in Islamabad, CIIT also has campuses at Lahore, Wah, Abbottabad and Attock. CIIT became functional in 1998. The success of the COMSATS Institute of Information Technology in Pakistan has inspired the establishment of a similar IT center at Damascus, Syria.

Academic Activities in Telecom and Microelectronics

1. CIIT offers undergraduate and post-graduate level degree programs in Telecom Engineering, Computer Engineering, Electronics and Applied Physics. Most programs offer courses in microelectronics and VLSI design. The institute has well equipped with Advanced VLSI design labs.
2. Department of Physics of CIIT has set up a Micro/Nano science laboratory with the collaboration of Higher Education Commission of the Government of Pakistan. The purpose of the laboratory is to produce manpower to take up the challenges of future in micro-electromechanical systems (MEMS) and nano-electromechanical systems (NEMS) allowing applications in areas ranging from bio-technology to air defense systems to chemical and biological diagnostics.

Establishment of the micro and nano-electronics laboratories at CIIT would help teaching and research programs in micro and nano electronics. We have scattered trained manpower in Pakistan, mostly located in R&D organizations in different cities of the country. There is a great demand at the university level for experts in this area. We would like to invite scientists from various parts of the country to come forward and help us in this set up. The laboratory set up at the COMSATS Institute of Information Technology established under the instant project will train students at the MS and PhD levels to meet this demand. It will cater the postdoctoral researcher, faculty and the students of the institute. These facilities are also available to all the scientists working in the other universities and R&D organizations in the country.

The lab facilities will be utilized to conduct research in the following areas:

1. Nanoelectronics (Micro and Nano Fabrication of Devices: Electronic and Photonic)
2. Synthesis and Characterization of Nanoparticles and Nanostructures (Nanomaterials) and their applications to environment, industrial processes, sensors, biology, etc.
3. Colloids and Organic Nanomaterials, Micro fluidics
4. Nano Biotechnology
5. Physics and Chemistry of surfaces
6. Nanocomputing/Computational Studies of Nanomaterials
7. Nanotechnology Business Management

13

Review of Electronic Industry in Sri Lanka

A.S. Pannila
Head/Technical Manager
Electro Technology Centre, Industrial Technology Institute, Sri Lanka
Phone: 0094-011-2691516; Fax: 0094-011-2691516
E-mail: pannila@iti.lk

ABSTRACT

In most of the South East Asian countries the electronic industry is one of the major economic activity. However, in Sri Lanka the electronic industry has not developed into a similar position compared to these countries. During the last 20 years the Government of Sri Lanka has made efforts to develop the industry. Technical assistance has been taken from several organizations to review and make recommendations to the government. The government institutions which are directly responsible for the development of electronic industry are : Board of Investment (BOI), Ministry of Industries (MOI), and Export Development Board (EDB).

There are about 25 electronic industries registered under MOI and another 25 under the BOI. The MOI industries are engaged mainly in the assembly of domestic items. The total output of these industries were estimated to about Rs 320 million in 2000 and provide direct employment for about 2,500 persons. Under the BOI, 25 projects have commenced commercial production. Out of these 11 are totally foreign owned, 5 joint ventures and 6 locally owned. The total investment for these industries have been estimated to Rs. 3,900 million and these have provided direct employment for nearly 10,000 persons.

The exports of the electronic products were not very significant compared to that from other sectors being only 2.6% of the entire industrial sector and 5.1% of the BOI industries. The corresponding figures for employment generation have been 1.6% and 3.4% respectively. Compared to the situation in other countries in the region these are very low and there is much scope for improvement.

Keywords: Electronic, Industry, Sector, Sri Lanka, Review, Status.

Introduction

During the last 20 years the Government of Sri Lanka has made efforts to develop the electronic industry. Technical assistance has been taken from several international organizations to review and make recommendation to the government. Out of all these studies, three studies given bellow are very important and their observation and recommendations are still valid. The author of this article also agree with the most of the observations and recommendations of these studies.

- Proposal for an Initiative for Electronic Manufacturing Sector-1999
- Master Plan Study on Industrialization and Investment Promotions - 2000
- Electronic Industry in Sri Lanka, status review and policy initiative for enhancement.

The electronic industry is a major economic activity in most of the South East Asian Countries providing employment to a large segment of population. However in Sri Lanka the electronic industry has not developed into a similar position. The Ministry of Industries and Board of Investment (BOI) has given priority to establish electronic industries under special concessionary terms.

The BOI has undertaken several studies with a view to capture the overseas market. Currently BOI is preparing a comprehensive marketing plan to promote the electronic industry as an export oriented industry.

The current status of the electronic industry in Sri Lanka revealed that about 25 industries are registered under the MOI and another 25 are registered under BOI. The total investment on BOI companies were estimated about Rs.5000 million in year 2002 and provide direct employment for about 11000 persons.

The three institutions, Arthur C Clarke Institute for Modern Technology(ACCMT), Industrial Technology Institute (ITI), National Engineering Research and Development Centre (NERD Centre) are the only places where the some R&D work in electronic is being carried out. These activities are limited to design of circuits to perform specific function, initiated as in-house projects on requirement from the industry.

Current Status of Electronic Industry in Sri Lanka

All the electronic industries operated in Sri Lanka has been registered under the following organization:

- **Ministry of Industries (MOI)**
- **Board of Investment (BOI)**

This chapter discuss the current status of the electronic enterprises operating under the above organizations.

Electronic Enterprises Operating under Ministry of Industries (MOI)

There are 25 electronic industries registered with the MOI and these industries receive benefits specially granted to the electronics sector. The main production of these industries could be categorized as home appliances and industrial products. The number of persons employed in these industries was approximately 2500 in the year 2000. In the same year the total output from these firms was estimated to be Rs. 320 million. All these industries sold their product in Sri Lanka and these industries could be further categorized as follows.

(*a*) Main production for the local market using assembling of imported kits.

(*b*) Specialized electronic products are designed and manufactured for local and export market.

Companies belonging to category (b) demonstrated innovative new product with new technology. Many of these companies had commenced business with limited capital, but were successfully able to venture into high tech areas in electronics and they demonstrated steady growth rate.

Potential for the Local Industry to Export

The Export Development Board (EDB) had surveyed potential export market for local electronic products. According to the findings of this survey EDB concluded that following product are potential for export market.

- Audio/visual system components
- Passive electronic components
- Printed Circuit Boards
- Radio and portable audio systems
- Colour TVs
- Personal Computers
- Bio-Medical equipment
- Musical instruments
- Telephone sets and PABX systems
- Home security systems
- Power supplies

The Export Development Board has been promoting the export of locally produced electronic product during last 10 years. This was achieved mainly by organizing overseas trade fairs, and meeting with potential foreign clients and consultants. However, there is no significant outcome of these exercise. Only a few passive components have been successful in finding steady export market. The main reason for this less export market for the Sri Lankan electronic product is they do not meet the desired standard required by export market. This is evident that few companies who consults the R&D organizations for quality improvement and assessment were able to find the export market.

Electronic Enterprises Operating under Board of Investment (BOI) of Sri Lanka

This section describes the current status of the export –oriented electronics industries established under the BOI. Twenty eight (28) projects have been approved by the BOI since 1983. The estimated investment for these projects were about Rs. 5,000 million. At present over 25 electronics companies are in commercial operation. These industries are located in Katunayake, Biyagama and Kandy Export Processing Zones. An overview of the electronic industries by project status in year 2002 is given in Table 13.2.

The growth of electronic sector has been very low compared with the other industries such as garments and related sectors. Even the number of electronic industries are small these companies involve large operation with 500 to 2000 employees per project. Most of these companies are produced magnetic components and power electronic related products. Two industries are involved with computer memory parts and they employ about 1500-2500 personnel. Table 13.3 shows that growth of BOI companies during the period 1995-1998 as statistics available at BOI. Figure 13.1 shows the increase of BOI electronic industries for the period of 1988-1999. Investment by type of ownership in year 1998 is given in Figure 13.2.

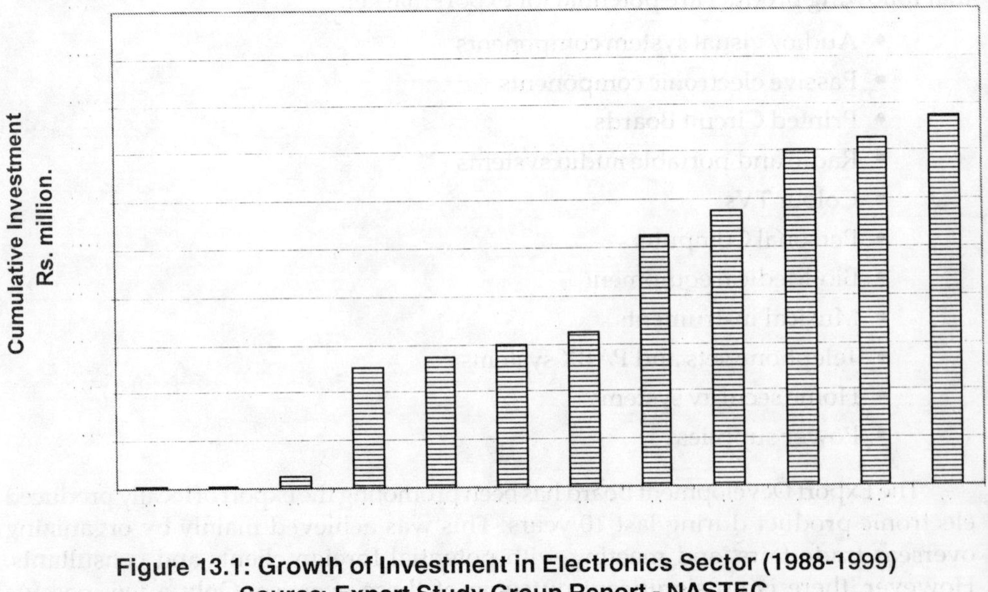

Figure 13.1: Growth of Investment in Electronics Sector (1988-1999)
***Source*: Expert Study Group Report - NASTEC.**

Table 13.1: Electronics Products Assembled /Manufactured In Sri Lanka (1999)

Product Type	*Annual Capacity No.*	*Value Rs. Million (1999)*	*Market*		*Process*	
			Local	*Export*	*Assem*	*Manu*
Television, Audio	39,000	73.3	X		A	
Television, Radio	47,553	12.6			A	
Television, Telephone	N.A.	38.2				
Telephone, Radio, Cassette	N.A.	NA			A	
Television	2,050	NA				
Television, Cassette	6,225	32.1	X		A	
Radio	1,200	NA			A	
Radio	1,500	NA			A	
Video Head	N.A.	NA			A	
Data Transmission Equip.	5,000	1.0	X			
Radio	10,500	NA	X		A	
Radio Cassette	30,000	9.8	X			
Bulbs	3,469,024	43.6	X			
Video, Cassette	3,000	NA				
Radio Cassette	4,000	5.5	X			
Electronic Item	N.A	NA			A	
Electronic Item	N.A	1.1			A	
TV Antenna	N.A	NA			A	
Electronic Item	4,000	NA				M
Electronic Circuits	N.A.	41.3	X			M
Electronic Mosquito Repellant	N.A.	NA				M
TV Antenna	N.A	NA				M

N.A.: Data Not Available; *Source* : MOI Database, 2001.

Table 13.2: An Overview of the Electronics Industry by Project Status – 2002

Project Status	*No. of Companies*			*Investment (Rs. million)*	*Employ-ment*
	Foreign	*Local*	*J/V*		
Commercial operation	11	6	5	3,871	9,994
Awaiting Construction	1	-	1	260	387
Awaiting Agreement	-	2	2	785	628
Total	**12**	**8**	**8**	**4,916**	**11,009**

Source: MOI Database, 2001.

Table 13.3: Electrical / Electronic Industry Projects with BOI Status and their Progress during 1995–1998

Company	*Colla-boration with*	*Employment Capacity*	*Products or Type of value addition*	*Export in Millions of Rupees (approximate figures only)*			
				1995	*1996*	*1997*	*1998*
Abans Jungpoo	Korea	543	Electrical/electronic Components	65	30	n/a	n/a
Aselco	Switzerland	181	Printed Circuit Boards	121	79	36	n/a
Asia 400	UK	51	Repair of used computers	1.6	1.5	1.9	4.4
Asia Prepress	Sri Lanka	38	Audio Products, Computers Software	0.4	3.8	0.3	–
Bartleet Micro-devices Lanka		1200+at time of suspension	–	325		37	Operation suspen-ded
Cortech Systems Lanka			–	59	62	1.2	0.1
Cruickshank & Partners Ceylon	UK	79	Voltage control panels	97	318	389	447
Elsuma	Switzerland	4.6	Electronic products (compact fluorescents)	190	122	57	773
FDK Lanka	Japan	1105	Magnetic heads, ferrite cores etc.	2025	4386	5469	5435
Flintec Transducers	Sri Lanka	20	Electric weighing systems	127	129	238	305
Frostaire Lanka	Sri Lanka	105	Air conditioners	12.6	12.2	1.3	–
IE International	Sri Lanka	6	Electrical and Electronics Goods	–	0.19	0.8	–
KIK Lanka	Korea	1098	Switchgear Assemblies	–	–	6	8
Kambool Maganetics	Korea	486	Magnetic Heads	26.2	87.7	407	135
Koolje Colombo	Germany	384	Crystal Units	204	220	229	195
Lanka Hiqu	Sri Lanka	420	Magnetic Heads	28.7	22.8	22.9	28
Nippon Maruichi Lanka Ele.	Korea	123	Microwave products	96	216	303	317
Shin Kwang Lanka	Japan	237	Fluorescent Bulbs	51	101	118	28

Source: Research and Documentation Department of BOI.

Investment and Ownership of BOI Electronic Industries

The total investment for electronic industry in year 1999 was Rs. 3800 million. It should be noted that this amount is only 19 million in year 1988 and 1500 in 1994 The statistics on investment in electronic industry indicate that about 91% investment from foreign and joint ventures type companies. The local investment is only 9%.

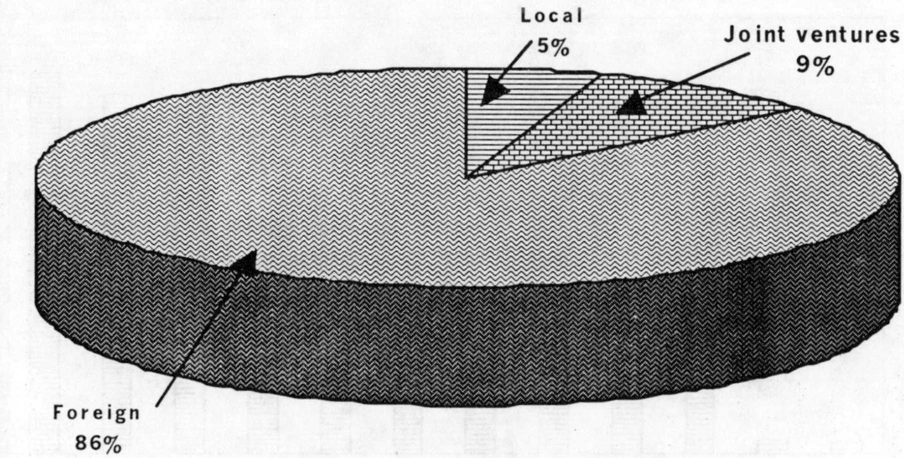

Figure 13.2: Investment by Type of Ownership (Electronics) – 1999
***Source*: Expert Study Group Report - NASTEC.**

Employment in Electronic Industry

The estimated employment in electronic industry is 10,000 including large number of graduate engineers. The average salary scales of the employees in the electronic sector is higher than that compared with other manufacturing industries. The statistic on employment reveals that 78% of the employees work in the foreign owned companies. Further the employment generation shows significant growth. It has increased from 384 in 1988 to 10,000 in 1999. Employment by type of ownership in electronic sector in year 1999 is given in Figure 13.3. Growth on employment in electronic sector is given in Figure 13.4.

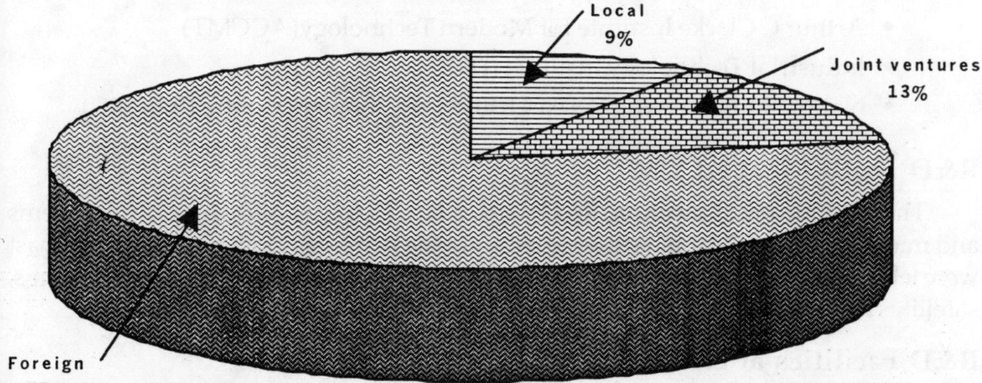

Figure 13.3: Employment by Type of Ownership – Electronics Sector (1999)
***Source*: Expert Study Group Report - NASTEC.**

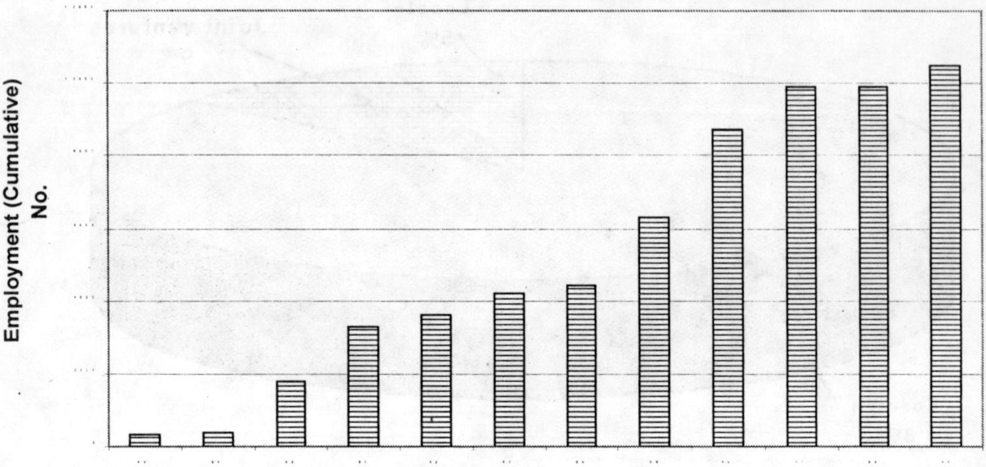

Figure 13.4: Growth in Employment Electronics Sector (1988-1999)
***Source*: Expert Study Group Report - NASTEC.**

Research & Development, Calibration and Testing Facilities for Electronic Industry

There is no single organization dedicated to the development of electronic in Sri Lanka. However, three national R&D institutes are partly involved in the R&D activities in electronic area. Development of R&D activities in electronic is very pour compared to the other areas. As such the areas of electronics does not get its due recognition or priority when resources are allocated. Hence it is necessary to have a separate institute dedicated to exclusively for electronic.

Three Institutes involved in the R&D activities in electronics are:

- Arthur C Clarke Institute for Modern Technology(ACCMT)
- Industrial Technology Institute (ITI)
- National Engineering Research and Development Centre (NERDC)

R&D Facilities at ACCMT

The institute has done more R&D work related to measurement control systems and micro-processor based hardware. Among the items developed in the recent past were telephone billing system, traffic control systems, UHF transceivers, PABX systems, satellites receivers etc.

R&D Facilities at NERDC

The Electrical & Electronic Engineering Development of the NERDC has been engaged on design and development of electronic equipment for several years and had successfully commercialized some of them.

R&D Facilities at ITI

The Electro Technology Centre (ETC) of ITI (Former Instrumentation Group) has developed many industrial measurement system on request made by clients. The ETC was formed in January, 2005 under re-structuring program of ITI. The main scope of ETC is to concentrate the R&D activities and other technical services needs to electronic industry. The main objectives of the ETC is given bellow.

- Design and development of electronic products
- Industrial processes automation
- Industrial measurement system automation
- Repair and maintenance of analytical equipment

One of the future objectives of the ETC is to develop the facilities to work as PCB design and sample center for country.

Calibration and Product Testing Services

Calibration Services

Any manufacturing industry has to use many measurement equipment for the quality control of their product. For apparel industry where a visual inspection could determine the general quality but any electronic product need test and measuring equipment. The measuring equipment and in the industry need to be calibrated by independent body. In such case these calibration laboratories have to maintain their tractability with specific international standards.

In Sri Lanka, Metrology Laboratories of ITI is the main institute who provide the industrial calibration with international tractability. The ITI metrology laboratory has the facilities to calibrate in the area of temperature, pressure, length, mass and electrical parameters. The temperature and mass calibration services of the ITI was accredited by SWEDAC in year 2004. Now they are going to extend their scope of accreditation to electrical parameters also. Electrical Calibration capability of ITI is given in the Table 13.4.

Table 13.4: Calibration Capability of ITI

Parameter	*Range*	*Uncertainty*	*Remarks*
DC Voltage	0-1,000V	±0.006%	Undertakes calibration of Voltmeters,
AC Voltage	0-1,000V	±0.06%	Ammeters, Multimeter, Wattmeter,
	1kV-40kV	±0.6%	Frequency Meters, Frequency
			Generators, Oscilloscope etc.
Direct Current	0-1.0 A	±0.02%	
	1.0A-10 A	±0.1%	
Alternating	0-1.0 A	±0.1%	
	10-50 A (40-400Hz)	±0.6%	

Testing of Electronic Products

The performance and quality of a product are two important factors that determine its survival in competition market. Industrialists need to build confidence among consumers for their products. Then only they could built their brand name. Any electronic product should be tested for their performance and quality before sale.

The central organization responsible for product testing is the Sri Lanka Standards Institute (SLSI). The SLSI has the limited facilities to testing of electrical/ electronic product. However, this is not adequate for the need of the industries. Therefore testing facilities for electrical/electronic product should be developed at one of the institution.

Problems and Constraints

The electronic industry is faced with many shortcomings, problems and constraints. To overcome the some of these constraints could be solved by correct involvement of R&D organization like ITI, SLSI, and ACCIMT.

- Production Constraints Encountered by Electronic Enterprises are given in Table 13.5.

Table 13.5: Production Constraints Encountered by Electronics Enterprises

Constraint	*Reason/s*	*Ranking*
High Production Cost	High Cost of imported raw materials and component inputs	1
	High air freight rates for electronic component	
	Labour cost is as high as neighboring countries	
	Stamp duty on import products	
Inadequate technology transfer	No special mechanism or incentives	2
	Lack of technology and know how through search and transfer	
Inadequate R&D to develop products	Lack of government focus on R&D	3
Poor Quality	Sub standard local components	4
	Lack of overall quality control (ISO standards)	
	Limited testing facilities	
Lack of trained manpower (*e.g.* Engineers, technicians, mechanics, chemists)	Education reforms are slow and not demand driven	5
	Vocational Training Programs designed	
	Lack of Quality Trainers for electronics industry	
Low level of support industries	Lack of vision and co-ordination at the highest level	6
	Not enough focus on "cluster approach"	

Source: Board of Investment of Sri Lanka – Survey 2000.

- Financial and Marketing Constraints Encountered by Electronic Enterprises are given in Table 13.6.
- The Constraints/Barriers identify by the Industrialists is given in Table 13.7.

Table 13.6: Financial and Marketing Constraints Encountered by Electronics Enterprises

Constraint	*Reason/s*	*Ranking*
Financial Constraint		
Limited access to commercial loans with low interest	Lack of government support (*e.g.* To create a special fund to provide low interest loans)	1
	Non availability of venture capital	
Lack of financial support to promote brand names/marketing	Lack of the government vision/recognition and assistance	2
Depreciation of Sri Lankan Rupee against major currencies	Government policies	2
Lack of financial assistance and tax allowances for foreign business related travel	Lack of government vision/recognition in supporting on "infant industry' like electronics enterprises in particular	2
Marketing Constraints		
Small Domestic market	Lack of confidence on domestic products over imported goods	1
	Relatively high cost and poor quality of domestic products	
	Limited availability of local/domestic products	
	Large inflow of imported goods	
	Lack of government procurement to encourage local production and sales	
Lack of a Sizable Export market	Inadequate information on foreign buyers	2
	Heavy leakage of duty free products to local market	
	Lack of strong links with established foreign industrial and trading companies	
	Lack of targeted products for the different segments of the export market	
Lack of confidence on available facilities (For data collection) and existing information	Lack of a cost-effective mechanism to provide timely information to enterprises at affordable or zero cost	3
Lack of a "help unit", a public private sector partnership (*e.g.* under the BOI) to look after the interest of the Electronics industry	No mechanism to solve bureaucratic, technological, financial or marketing constraints(short-medium & long-term faced by the Electronics industry)	4

Source: Study Report on Electronic Industry in Sri Lanka – NASTEC.

Table 13.7: Constraints and Barriers Identified by Industrialists

Sl.No.	Constraint/Barrier	Percentage who responded	
		Positive	Negative
1.	Inappropriate tariff structure, especially with regard to import of raw materials Vs import of finished goods.	57	43
2.	Lack of incentives and concessions for the export of products manufactured locally	28	72
3.	Lack of facilities for instrument calibration and product testing for quality assurance	71	29
4.	Difficulty in recruiting engineers competent to undertake design of circuits and products	43	57
5.	Lack of link industries to supply necessary components such as attractive casings, pcbs, metallic fittings, labeling etc.	86	14
6.	Lack of access to data on new devices and components that are necessary for product development	71	29
7.	Lack of access to new technologies and foreign collaborators to undertake joint ventures	57	43
8.	Difficulty in identifying and contacting overseas customers to market the products	57	43
9.	Difficulty in raising bank loans for developing new products based on new technologies	86	14
10.	Lack of training facilities for engineering and technical staff to undertake development of new products and manufacture	57	43

Source: Study Report on Electronic Industry in Sri Lanka – NASTEC.

References

- BOI study on electronic industry in Sri Lanka-1998/2000.
- Electronic Industry in Sri Lanka, Status Review and Policy Initiatives for Enhancement – Study Report – NASTEC, 2002.
- Kularatne N, (1998). Proposal for an Initiative for the Electronic Manufacturing Sector.
- Master Plan study on Industrialization and Investment Promotion, Study Report –MOI & JICA –2000.

Abbreviations

ACCIMT: Arthur C Clarke Institute for Modern Technologies

ADB: Asian Development Board

BOI: Board of Investment

CISIR: Ceylon Institute of Scientific and Industrial Research

EDB: Export Development Board

ETC: Electro Technology Centre

FTZ: Free Technical Zone

IDB: Industrial Development Board

ISO: International Standards Organization

ITI: Industrial Technology Institute

JICA: Japan International Cooperation Agency

MOI: Ministry of Industries

MOST: Ministry of Science & Technology

MUSSD: Measurement Units Standards and Services Department

NASTEC: National Science and Technology Commission

NERDC: National Engineering Research and Development Centre

PCB: Printed Circuit Boards

R&D: Research and Development

SLSI: Sri Lanka Standards Institute

14

Status of Microelectronics Development in Tanzania: An Overview of Legislative Frameworks

Emmanuel J. Kaale
Scientific Officer,
Tanzania Commission for Science and Technology, Tanzania
E-mail: ejkaale@costech.or.tz

Introduction

In the context of this paper, Microelectronics is taken as a subfield of electronics. Microelectronics, as the name suggests is related to the study and manufacture of electronic components which are very small. These devices are made from semiconductors using a process known as photolithography. Many components of normal electronic design are available in microelectronic equivalent: transistors, capacitors, inductors, resistors, diodes and of course insulators and conductors can all be found in microelectronic devices.

The world is experiencing the information revolution through rapid transformation of technology. The engine of the growth has been through miniaturization or microelectronics. There is no doubt therefore that "microchips (integrated circuits)" which are the hearts of all electronic equipment play a crucial role in this transformation. Microchips with combinations of software are determined by the "ability" of all the IT equipment and the nature of the IT industry and society. Invention of integrated circuits and their continuous progress have made the world's information era possible and it will continue to revolutionize the world. Demand for microchips is growing continuously at exceptionally high annual rates and there is still no sign of saturation thanks to the present worldwide growth of the Internet, telecommunications, and other coming-up yet-unthinkable applications. In fact when we say electronics is one of the fastest growing areas in research and industrial development we refer indirectly to what is happening in the microelectronics area. Because of the cross cutting nature and near universal application Microelectronics is being accorded strategic importance in Tanzania and has been identified as one of the areas of priority in both research and industrialization.

Tanzania Background

Tanzania is an East African Country with 945,000 sq. km and a climate that ranges from tropical to arid to temperate. The present population is 30 million. Two major languages are prominent Kiswahili (official), and English.

The Social Sector

Education: Attendance–74% (primary).

Literacy–67%.

Health: Infant mortality rate–98/1,000.

Life expectancy–45 years.

Work force: Agriculture–80%.

Industry, commerce, government–20%.

The Economy GDP (1997): $7.5 billion.

Annual growth rate: 3.6%.

Per capita income: $260.

Natural resources: Hydroelectric potential, coal, iron, gemstone, gold, natural gas, nickel, and diamonds.

Agriculture (60% of GDP): Products–coffee, cotton, tea, tobacco, cloves, sisal, cashew nuts, and maize.

Industry (10% of GDP): Types–textiles, agribusiness, light manufacturing, oil refining, construction.

Trade: Exports–$793 million: coffee, cotton, tea, sisal, diamonds, cashew nuts, tobacco, and cloves. Major markets–U.K., Germany, India, Japan, Italy, and Far East. Imports–$1.2 billion: petroleum, consumer goods, machinery and transport equipment used clothing, chemicals and pharmaceuticals. Major suppliers–U.K., Germany, Japan, India, Italy, U.S. The is a lot to be done in the industrialization of Tanzania that is why microelectronics holds the key to transformation.

Legislative Frameworks

Because of the appreciation of the role of microelectronics to national development, Tanzania has resolved to implement an effective and sustainable microelectronics programme. The key steps include the establishment of the basic legislation that will enable the country to proceed along unhindered. Such legislation includes the National

Science and Technology Policy, the ICT Policy, Energy Policy and Telecommunications Policy. The appropriate regulatory mechanisms have also been set and this demonstrated through the merger of the former Tanzania Broadcasting Corporation and Tanzania Communications Commission into the now Tanzania Communication and Regulatory Authority. Related authorities established recently include Surface and Marine Transport Regulatory Authority (SUMATRA) and the Tanzania Civil Aviation Authority.

A Determined National Commitment

The Tanzania government is committed to support and promote the microelectronics industry through setting up incubation centers to attain higher value added and technology based microelectronics products. A Technical Committee composed of the Ministry of Communication and Transport, University of Dar es Salaam, Tanzania Commission for Science and Technology; Ministry of Science, Technology and Environment; and the private sector is working on developing an implementation strategy for the ICT policy.

Tanzania has cooperated with the UNECA to set these legislative frameworks which besides creating sensitization among the decision makers has also brought about change of heart on the part of government towards lifting a 1974 ban on the importation of computers, appointment of a national ICT coordinator, undertaken e-Readiness assessments and supported the introduction of various ICT based curricula for all schools staring from primary one.

The components of an informatics policy are therefore fully covered in the ICT policy which in the context of this paper refers to a plan for the development and optimal utilization of information technology, data resources and services. Due to these efforts the potential benefits of informatics in national development the level of informatics policy activity in Tanzania has grown tremendously. Once the policies are in place it becomes necessary to analyzes them and to assess the extent to which they achieved their objectives. The country has recently approved the set up of a Technical Committee to development implementation strategies of the ICT policy.

Zwangobani, (1995), The main objective of national informatics policy is to cost-effectively acquire and optimally use informatics to process information as an integral factor in all sectors of national development. The possession and use of knowledge are, without doubt, essential ingredients for national progress. Information, the communicable form of knowledge, especially scientific and technological information, is one of the main prerequisites for economic and social development.

Zwangobani, (1987) further testifies that most countries in Africa have sectoral IT policies governing the implementation and operation of IT programmes. Some are micropolicies which operate at the organization level while others are macropolicies which are applicable at the national level. The IT policy in Tanzania ias one example of this and has the following components:

- The acquisition and/or development of the technology
- The development of human resources and their use including informatics in the academic system

- The establishment of a telecommunications infrastructure
- The storage and use of data in private and public sector databases
- The protection of intellectual property rights
- Participation in extra-national information networks and exchange.

This paper discusses Microelectronics activities in Tanzania in the following areas: Hardware, Software, Education and Training, Data Banks, Telecommunications, Research and Development and Consultancy.

Microelectronics Infrastructure and Environment

From the industrial point of view, Tanzania still lacks a strong electronics manufacturing base, has not attained the required level of competitive and good quality engineers, technicians and managers with a long-term supply potential although the numbers of universities and student enrollment are on the fast growth track. Tanzania also lacks fast growing number of IC design companies and Hi-tech business supporting environment though the intention is to set up incubation centers, pilot plants, training centers and financial liberalization services. Further the Tanzania has the most Competitive investment incentives in Africa south of the Sahara. The scheme includes exemption of import duty on machinery and raw materials, and tax holidays. The infrastructural support sector is fast growing and logistics with competitive costs. High quality industrial estates with all necessary infrastructures are available. Though currently not adequately available, Tanzania has the potential to provide enough electricity, water, land and logistics with very competitive cost performance. Tanzania is surrounded by an existing large and fast growing markets with surrounding high growth developing countries of over 100 millions population and good relations with all its neighbors.

Education and Training in Microelectronics

The microelectronics industry in Tanzania is at its infant stages. Through the establishment of extensive IT based training programmes soon the necessary human resource base will be established to support the microelectronics industry. The Institute of Curriculum Development of the Ministry of Education and Culture has developed and recommended a syllabus for computer education in schools. An intervention is in its final stages to introduce computer training courses in all 39 Teachers Colleges in Tanzania as the first step towards introducing computer courses in schools.

The University of Dar es Salaam has succeeded in offering a BSc (with Computer Science) and post graduate degrees, IT teaching staff recruitment and retention, improved supply of teaching materials, text books and other facilities. The University has since established a computer centre and is said to be one of the most networked in Africa except South Africa.

The Dar es Salaam Technical College has been transformed into a higher Institute of Technology (DIT) and is one of the key institutions in the country offering computer engineering courses. It incorporates microcomputer and hardware maintenance into its diploma, full technician certificate and advanced diploma courses. The Institute of Finance Management (IFM) offers a range of informatics courses integrated with their diploma courses in Accountancy and Banking besides the specific IT ordinary and higher diplomas. In addition the Institute also offers short management information programmes. A number of other public sector institutions also offer informatics programmes; these include the Mzumbe University, the Cooperative College, the College of Business Education, and the Schools of Accountancy in Dar and Arusha. The St Joseph University of Computer Engineering is a private initiative and is one of the most specialized higher learning institutions in the Information Technology area.

Microelectronics and Related Information Appliances

Information appliances include appliances such as radio sets; computers; telephone sets; television sets; fax machines; and personal communicators. These are the devices people use to generate, send, and receive voice messages, data, video, and multimedia information, as well as simply communicate with one another.

According to World Bank Report 1999/2000 Tanzania is reported as having a total of 278 radios per a thousand people. Data available for other appliances per a thousand populations include: 21 TV sets; 1 mobile phones (approximately 35,000 sets in the country); 1.6 computers; and an average of 1.4 Internet hosts per ten thousand persons. The introduction of about 1200 payphones has considerably increased access terminals to end-users. Following liberalization of trade in Tanzania in mid 1980's; there have been a considerable increase in the number of information appliances in the country. The Government has reduced taxes on appliances such as computers and telecommunications equipment imported into the country.

Microelectronics and Telecommunication

Communications networks comprise telecommunication networks including local exchanges, Internet Service Providers; electronic media, print media, local and wide area networks; cellular; cable TV providers; alternate access vendors; TV and radio broadcasters; and long distance networks, including terrestrial and satellite.

Telecommunications as a Catalyst to Development

Tanzania is taking up the challenge of developing, from modest beginnings, a modern telecommunications sector capable of supporting broad-based national economic and social development. As a result of telecommunication sector reform, which began in 1993, Tanzania is finally overhauling its dilapidated telecommunications network. The sector reform which included a five-year project (TRP) to improve the country's telecommunications network is eventually paying off. Digital exchanges and microwave technology have connected thousands of new subscribers. Switching from analogue to digital technology has increased the capacity, reliability and quality of Telecommunications. The overall telephone coverage of the country has improved at the same time from 1.5 per cent of the population to 2.2 per cent and is expected to rise to 2.8 per cent by the end of 2000. Improvement of telecommunication infrastructure has resulted in improved overall information access.

Internet Services

There are currently 14 Internet Service Providers (ISP) in Tanzania. Competition has resulted in lower charges (from USD200/month to USD50/month) and increase in quality and number of Internet services. Most of ISPs and other users strongly feel that prices could further go down if more data carriers could in the existing Internet Exchange point. However availability of bandwidth remains a serious constraint.

Currently ISP's are only regulated on possessing communications equipments but not what is transmitted. Laws exist against pornography, seduction (political or sexual) and provocation and right to beliefs.

There are a number of on-line services that have been introduced in Tanzania. Besides provision of common basic services, such as e-mail, the World Wide Web, newsgroups, web hosting, web-authoring and other value added services like Internet fax some ISPs already had unofficially started to introduce VOIP applications even before it became officially authorised early this year.

In Tanzania S & T comes under the National Commission for Science and Technology. Membership is from the science and technology system, the productive sector, public sector, technical ministries etc. It is charged with the coordination of S & T planning as well as policy implementation. Its main function is to act as a stimulant and catalyst for the development of indigenous technologies.

In addition to the establishment of the Commission for Science and Technology, a National Centre for the Transfer, Adaptation and Development of Technology was also to be set up. Such a centre was to serve as a mechanism for the assessment and choice of imported technologies on a systematic basis. Among the core functions of the Centre are:

- Acquire and analyze information on alternative sources of technology and its delivery to users
- Act as a registry for imported technology and to keep register of the domestic technological resources and manpower.

The use of microelectronics products and process technologies has definite impact not only at the development stage of an economy but also can influence the direction of that development at later stages. A country's domestic environment and international involvement determine how far and in what way it can exploit potentials of those new technologies.

Conclusion

The importance of an efficient telecommunications infrastructure to the development of the informatics sector cannot be overemphasized. Efficient data communication infrastructure in Tanzania is simply hard to come by.

References

- Synthesis of 5 Anglophone Africa Country Studies on the Effectiveness of Informatics Policy Instruments in Africa by E. Zwangobani, October 1995

- United Republic of Tanzania, (1985). The National Policy Science and Technology Policy for Tanzania.
- Zwangobani, E. (1988). National Informatics Policies in Sub-Sahara Africa, IDRC Report and Proceedings of a Regional Seminar, Addis Ababa, Ethiopia, 28 Nov. - 1 Dec., 151-174.
- http://www.tcra.go.tz/Publications/nationa % 20ICT % 20Policy % 20of % 202003.pdf
- http://www.costech.or.tz/contact%20address.htm

15

Status of Microelectronics in Mauritius

Bharati Kisto
Information and Communication Technologies, Mauritius
E-mail: bkisto@imail.icta.mu

ABSTRACT

The last ten years have witnessed enormous progress in wireless communication. This progress is mainly due to the technological advances in microelectronics. These advances in technology have made telecommunication services accessible and affordable to a greater number of users especially in the developing world.

No new wireless technology may however, emerge unless the appropriate regulatory framework is present in order to support its emergence. The policy maker and regulators have had to adapt their regulations and policies in order to provide these technologies with the necessary resources, especially in terms of radio frequency spectrum.

Technology is evolving rapidly. However the regulatory framework is finding it more and more difficult to cope with this advancement. Regulatory agencies, especially in the developing countries are facing challenges to adapt to these fast-paced changes in wireless technology.

This article presents some of the regulatory issues considered in Mauritius in order to facilitate the introduction of technologies such as 3G mobile and considers some of the challenges which the regulatory framework in Mauritius and other developing countries may face as a result of the ambition of the industry to introduce and use emerging wireless technologies.

Introduction

Microelectronics has been a significant factor in the emergence of several new technologies, including in wireless communication. Mauritius has recently witnessed the introduction of several such new technologies such as Third Generation Mobile Technologies (UMTS and CDMA2000), broadband access technologies for example Wi Fi and Wi Max, and Radio Frequency Identification (RFID). The existing regulatory provisions had to be revamped in order to create a conducive environment for their introduction. This paper presents the various regulatory issues that were considered in the facilitation of the introduction of these technologies.

Whenever a new technology is introduced for wireless communication, it is very important that the regulatory framework adapts itself in order to:

- Ensure that resources such as frequency spectrum are adequately available for the operation of the new equipment and systems and;
- Ensure that operation of these equipment do not cause harmful interference, in any way, to the operation of existing systems and do not constitute health hazards.

Measures such as frequency coordination, migration of existing systems if required, and development and determination of standards specific to these new technologies have to undertaken by the National Regulatory Agency in order to facilitate the introduction of new technologies while at the same time protecting existing radio communication systems.

It is worth noting that advances in wireless technologies, driven by the development in microelectronics, are completely reshaping the current radio communications model. The legislation and regulatory framework in most developing countries, including Mauritius, are finding it increasingly difficult to adapt to the challenges posed by these technological advances.

This paper provides an overview of the regulatory framework with respect to wireless communication and covers:

- RF technologies and Radio Spectrum management issues;
- RF Technologies and standardization issues;
- The type approval procedure adopted by the regulator in order to ensure the any new type of radio communication equipment is according to specified standards.
- The regulatory framework in Mauritius pertaining to the introduction of new technologies;

ICT Sector and Regulatory Framework in Mauritius

The Information and Communication Technologies (ICT) sector in Mauritius was fully liberalized in 2003. Prior to liberalization, new legislations were adopted and a National Regulatory Agency, the Information and Communication Technologies Authority, was established.

The radio communication sector in Mauritius is also regulated under the Information and Communication Technologies Act 2001 (as amended).

Role and Functions of the National Regulatory Agency

The Information and Communication Technologies Authority has been established under the Information and Communication Technologies Act 2001 and is mandated to, among others, license and regulate the information and communication service and to further the advancement of technology, research and development relating to information and communication technologies.

The ICT Authority has, inter alia, the functions to:

- Ensure the safety and quality of every information and communication services including telecommunication service and, for that purpose, determine technical standards for telecommunication network, the connection of customer equipment to telecommunication networks;
- Allocate frequencies and manage, review, and, where appropriate, reorganise the frequency spectrum;

The International Telecommunications Union

Mauritius has acceded to the Constitution and Conventions of the International Telecommunications Union in 1968, shortly after gaining independence. Therefore, as a Member State of the International Telecommunications Union, Mauritius has the obligation to abide by the regulations of the ITU, including the Radio Regulations which addresses issues pertaining to radio communications.

Radio Spectrum Management Issues and New Technologies

Because two or more radio signals occurring simultaneously and over the same frequency can interfere with each other and nullify their benefits, spectrum must be managed to prevent interference (Lie, E., 2004).

Radio frequency spectrum is a limited resource and must be used in the most effective manner possible. With its low cost and ability for rapid deployment, wireless communications is rapidly becoming the preferred infrastructure solution for developing countries both for voice and data transmission. This has in turn increased the demand for radio spectrum resources. As a result, regulators are under constant pressure to allocate this scarce resource evenly among competing users.

The radio frequency spectrum in Mauritius is allocated to the various radio communication services based on the allocations inscribed in the Radio Regulations of the International Telecommunications Union. The Mauritian radio frequency allocation chart is presented at Figure 15.1.

No new wireless technology may emerge unless the required frequency spectrum is made available for the operation of equipment and systems based on these technologies. The introduction of new technologies requires the regulator to review the radio frequency allocation plan as well as the band plan from time to time and to implement measures such as spectrum re-farming.

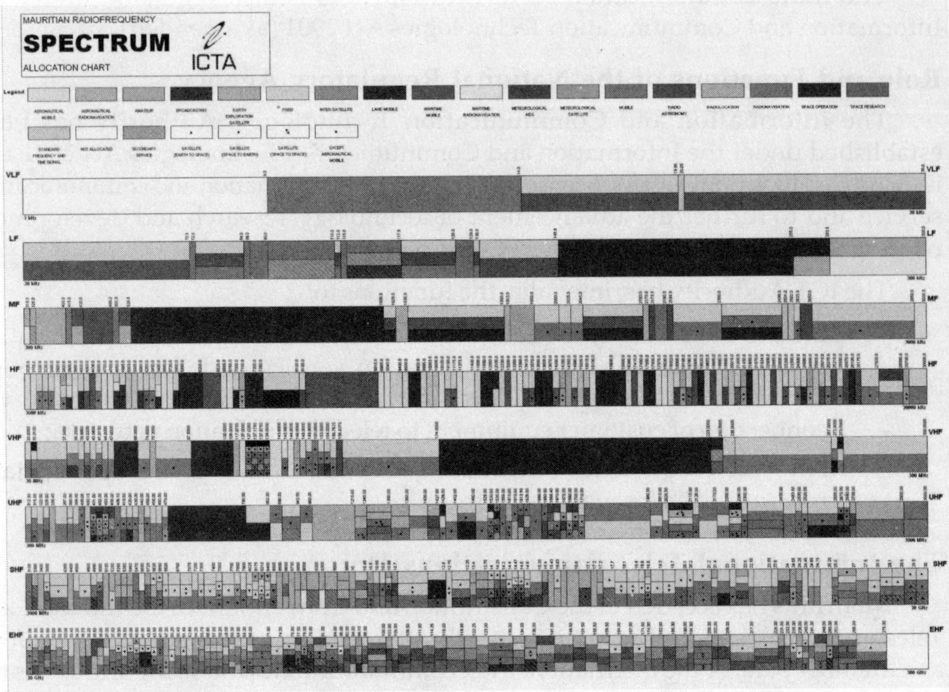

Figure 15.1: The Mauritian Radio Frequency Allocation Chart

Standardization Issues

In addition to allocating radio spectrum to these new wireless technologies, the National Regulatory Agency also has to develop and determine standards specific to these new technologies in order to facilitate their introduction while at the same time protecting existing radio communication systems.

The Importance of Standards

The regulator in Mauritius recognizes that adoption of international standards is beneficial for society at large including governments, businesses and consumers. Standards play an important role in raising levels of quality, safety, reliability, efficiency and interoperability – as well as in providing such benefits at an economical cost. Standards contribute towards making the development, manufacturing and supply of products and services more efficient and safer. Standards provide governments with a technical base for health, safety and environmental legislation. They aid in transferring technology to developing countries. They also serve to safeguard consumers, and users in general, of products and services - as well as to make their lives simpler.

Type Approval Procedure

The telecommunications regulator in Mauritius, the Information and Communication Technologies Authority has as one of its functions to under the law

'Ensure the safety and quality of every information and communication services including telecommunication service and, for that purpose, determine technical standards for telecommunication network, the connection of customer equipment to telecommunication networks'. The regulator fulfills this function by requiring all equipment to be used in Mauritius to be type approved by it.

Type approval procedure is the procedure by which the regulator determines whether an equipment may be used within the territory of Mauritius.

The main goal of the type approval procedure is to ensure that all radio communication equipment that are used in Mauritius comply with international standards approved by the Authority. Type approval of equipment also ensures that no sub – standard equipment which can represent health and safety hazards are operated in Mauritius.

Type approval further ensures that the operating frequency of all radio communication equipment is as per the frequency spectrum allocation in Mauritius and that no interference is caused to essential services.

In order to be type approved by the regulator, the equipment must comply with specified standards. Wireless equipment is required to comply with standards with respect to:

- Safety requirements, that is, the piece of equipment in no way causes any harm to the user or to any other person or system and;
- Electromagnetic compatibility, that is, the equipment or system is able to function satisfactorily in an electromagnetic environment without introducing intolerable electromagnetic disturbances to anything in that environment.

In addition the equipment must be constructed in such a way that it effectively uses the spectrum allocated for its operation without emitting intolerable levels of spurious emissions.

The type approval procedure followed by the regulator in Mauritius in order to ensure that equipment based on existing and new technologies are according to specified standards, is described in Figure 15.2.

New Wireless Technologies and Regulatory Framework in Mauritius

Mauritius has recently witnessed the introduction of two major wireless technologies which are the result of recent advances in microelectronics. While the Third Generation (3G) mobile service was launched at the end of 2004, Wi Max is being currently launched in Mauritius. Mauritius was in fact the first African country where 3G mobile services were offered.

The introduction of both these technologies has necessitated much planning by the regulator. The regulator has gone through several steps before allocating spectrum for these wireless technologies.

Step 1 The formulation of the proposals by the regulator after determining the resources that would be required in order to enable the introduction of the

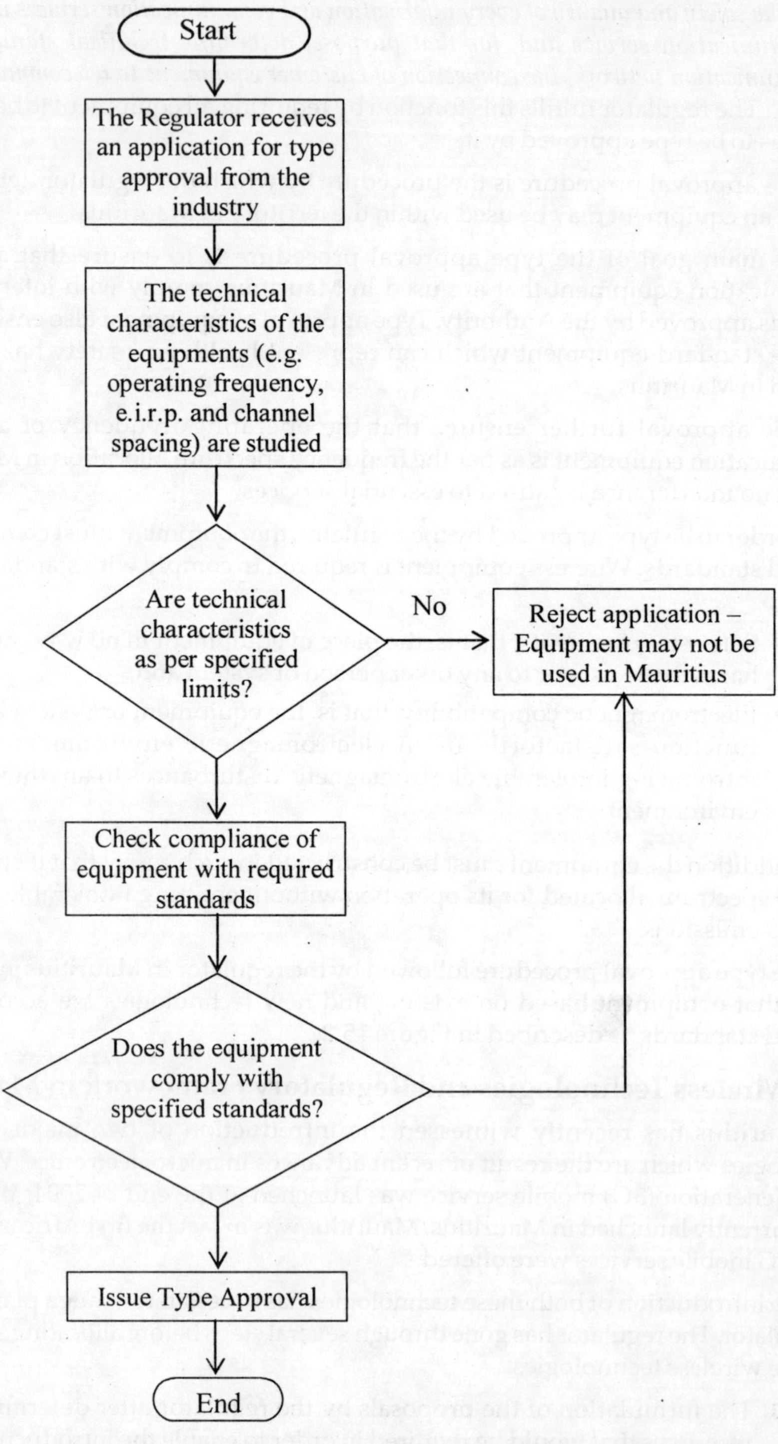

Figure 15.2: Type Approval Procedure Followed in Mauritius

new technology. This includes the review and re-organisation of the frequency spectrum and where necessary, devising migration plans, in the event that existing systems were to be moved from the required frequency bands. The technical characteristics and standards with which the equipment and systems based on these new technologies would need to comply were determined.

Step 2 The carrying out of a public consultation exercise where the proposals of the regulator were presented to various stakeholders and their views were gathered.

Step 3 The representations made were compiled by the regulator and presented to all stakeholders in on open house session. The views of these stakeholders were further gathered during the Open House session.

Step 4 Decision of the Authority were made public and licences were issued to operators.

Both in the case of 3G mobile and broadband wireless access technologies, the regulator in Mauritius has responded favorably to requests made by the industry and developed the necessary framework for the introduction of these technologies.

Advances in RF Technology and Challenges to Regulators

The currently RF regulatory framework of Mauritius and several other developing countries, is based on traditional communications systems whereby communication may occur only when the receiver senses a signal stronger than the background noise, at predefined frequency, allocated exclusively to one user. This traditional communications model is constantly being challenged by technological advances in the wireless communication.

One such wireless technology which has disrupted the radio communication and radio spectrum allocation model is Wi Fi, an application of spread spectrum technology. Wi Fi does not require that the frequency band in which it operates be allocated exclusively to one user. Wi Fi enables users to share the 2.400 – 2.4835 GHz frequency band without causing any interference to one another.

Advances in RF Technology

Other wireless technologies which defy the traditional communications model are emerging rapidly, driven by advances in the field of microelectronics. Examples are Ultra Wide Band Transmission (UWB), smart antennas, and Software Defined radios (SDR).

Ultra-Wide band is one of the most promising radio frequency technologies because it can transmit data at very high speeds by sending the transmission over a wide range of frequencies at very low power levels (Reynolds, T. 2004). They thus enable the operation of more than one system in the same frequency band without causing any interference to narrowband systems which may be operating in the same band.

Another way of improving the efficiency of a communication system is the use of better antennas. Antennas being used currently are passive and 'dumb' devices.

However, with the advances in microelectronics, smart antennas are being developed. These devices are intelligent enough to determine the source of the transmission. A digital processor does the thinking and analysis of the RF signals that give the unit its intelligence (Reynolds, T. 2004).

The radios being developed are also becoming 'smarter' with respect to spectrum usage. Software defined radios (SDR) are adaptable to different circumstances. The technical characteristics of these radios are software defined and may be programmed to accommodate to themselves to different operating environments. One specific type of SDR is the agile radio which may operate on any unused frequency on which it does not sense any activity. These radios may operate over a wide frequency spectrum but do not require any frequency allocation being made expressly to them.

Regulatory Challenges for Developing Countries

While these new technologies promise to make more efficient use of radio spectrum, they also represent serious challenges for regulators and policy makers. Regulators would need to adapt to these changes, which is not an easy task for those like the Mauritian regulatory agency, which have been recently established and are still in the process of building themselves.

The biggest challenge is to keep pace with the rapidly developing technological advances and develop standards fast enough in order to regulate the operation of systems and equipment based on these new technologies, ensure interoperability between different systems and also protect existing system.

Another challenge would be to define policies with respect to spectrum allocation and licensing. In most countries spectrum is allocated to licensees by the regulators and this constitutes a major source of revenue for the regulators. For instance these regulators would need to review their policies to account for software defined radios such as agile radios which do not require any spectrum allocation.

Conclusion

Microelectronics has significantly contributed to the development of wireless technologies and played a very important role in making telecommunication services accessible to a larger number of users as a result of the low cost and ability to rapidly deploy wireless systems.

The introduction of new technologies such as Third Generation mobile technologies and broadband access technologies in Mauritius has necessitated that the regulatory framework in Mauritius be revamped.

This paper has presented the regulatory aspect of the application of microelectronics for wireless communications in Mauritius and covered the various regulatory issues considered in the country in order to facilitate introduction of new technologies.

Also emerging wireless technologies, driven by the development in microelectronics, are posing challenges to the regulatory framework in Mauritius. Other regulatory agencies in other developing countries would, no doubt, be facing similar problems.

These challenges and potential problems may need to be considered during the development stage of the wireless technologies. This would undoubtedly facilitate their introduction in developing countries.

References

- Efstathiou, D., Fridman, J, Zvonar, Z., (1999). Recent Developments in Enabling Technologies for Software Defined Radio, IEE Communications Magazine
- International Telecommunications Union Radio Regulations, (2001). Geneva Switzerland.
- Lie, E., (2004). Radio Spectrum Management for a Converging World. Background paper – ITU Workshop on Radio Spectrum Management for a Converging World, Geneva, Switzerland.
- Reynolds, R., (2004). Advanced Wireless Technologies and Spectrum Management. Background paper – ITU Workshop on radio Spectrum Management for a Converging World, Geneva, Switzerland.

16

Status of Microelectronics in Egypt

Hamed Elsimary
Chairman,
Electronics Research Institute, Microelectronics Department,
El Dokki, Cairo, Egypt
E-mail: hamed@eri.sci.eg

Egypt and the Information Society

The formation of the new government under Dr. Ahmad Nazif revived hopes for economic growth and for the creation of an information society, one that can take Egypt into a new technological era, one that can bring it up to date with current technological development in a world where the Information and Communication Technology (ICT) sector outperforms other conventional sectors of the economy. Does Egypt have an information society that can boost economic growth? And if such a society exists, even tentatively, how can we develop it further? The new Prime Minister, with his ICT background, must be particularly aware of the requisites of the free market and economic globalization.

Where Does Egypt Stand on the Information Society?

One of the new government's priorities is to implement the e-government project, as part of President Mubarak's initiative to turn Egypt into an information society. Earlier, efforts have been made to prepare Egypt for the digital age, alleviate red-tape, reform taxes, attract investment, and build a strong ICT industry focusing on exports.

As part of Egypt's bid to complete the legislative work facilitating its entry into the ICT age, the country has passed laws concerning intellectual property, e-commerce, consumer protection, computer piracy, and e-signature. The above measures were taken in response to the growth of e-commerce worldwide. E-commerce has posed a challenge for governments across the globe, forcing the latter to provide the infrastructure suitable for ICT growth as well as the legal framework providing maximum security for e-transactions.

Egypt has come up with an initiative to create an information society and a strong, export-oriented ICT industry. The initiative rests on seven points:

1. Digital preparedness, which involves providing ICT services to all citizens across the country.
2. E-government, which involves providing government services to citizens and investors wherever they are and at the required speed.
3. Electronic infrastructure, which involves building an information society, a society that can function with less hardcopy and banknotes.
4. E-education, which involves raising the competitive capabilities of the Egyptian workforce. The education and higher education ministries are actively introducing ICT in curricula, an effort which involves the use of interactive systems.
5. Digitalization of the health sector, which is conducted in cooperation with the Ministry of Health.
6. Digital documentation of the country's cultural legacy and natural wealth. The Center for Documenting Cultural and Natural Heritage has been set up in cooperation with the Bibliotheca Alexandrina.
7. Promotion of the ICT industry with the aim of increasing Egyptian ICT exports and motivating global ICT companies to base their African and Middle East operations in Egypt.

Challenges and the Prospects of Growth in Egypt

In a study prepared for the Davos World Economic Forum about the challenges of ICT development in the Arab world, Arab countries were divided into three categories: a fast growth group, including Kuwait and the United Arab Emirates; an emerging group, including Egypt, Jordan, Lebanon, and Saudi Arabia, and a starter group, including Morocco, Oman, and Syria.

The digital preparedness index gauges the digital business environment, the viability of ICT infrastructure, government computerization, and the size of e-commerce in various countries. In 2003, a list was published of countries gauged by that index. The list includes 60 countries, only three of which are Arab: Saudi Arabia (ranked 45), Egypt (ranked 51), and Algeria (ranked 58).

Egypt has made immense effort to provide the right environment for a scientific and technological breakthrough, but time is needed for such effort to pay off, and there is need for further steps to be taken. Information society does not require an intricate web of legal measures, but it calls for a full awareness of the goal and structure of that society.

Information society requires a strong infrastructure and legal environment, training systems for personnel, improved education, and incentives to local and foreign investment. It requires freedom and transparency in the transfer and use of data, as well as a democratic climate allowing for the type of diversity that encourages innovation. The new government needs to rid the domestic electronics industry from its endemic problems. Most of these problems are not as related to scientific and technical inadequacy or the shortage of personnel as they are to managerial shortcomings and the lack of vision. The government should support local industry, improve education and electronics curricula, and form a national agency to co-ordinate among various services.

The e-government can materialize through appropriate technical solutions guaranteeing the probity of government services, and the integration of domestic information systems with foreign agencies. This going to be particularly important in the light of the expected growth of e-commerce over the next few years, with the signing of GATT and the increased use of the internet.

Monthly Indicators June 2005

Communications

Telecom Egypt

- Capacity of local public switching exchanges (12.3 Million Lines)
- Telephone main lines in operation (9.9 Million Lines)
- Waiting list for main lines (48798 Thousands)
- Fixed Lines Penetration (14%)
- International Circuites (15.292 Thousands)

Number of Companies in CIT Sector

Mobile Subscribers

Total Cellular Mobile Telephone Subscribers ***(9.400251 Million)***

- Cellular mobile telephone subscribers - Mobinil (5.074742 Million)
- Cellular mobile telephone subscribers - Vodafone (4.325509 Million)
- Mobile Penetration (9.95%)

Public Pay Phones

Total Public Pay Phones (55227 Cabins)

- Public pay phones - Menatel (30810 Cabins)
- Public pay phones - Nile Telecom (18618 Cabins)
- Public pay phones - Telecom Egypt (5799 Cabins)

Internet

- International internet band width (2.88 Gbps)
- Internet Users (4.4 Million)

Communication and Information Technology

- Number of established Companies Information Technology (1382 Companies)
- Communication (67 Companies)
- Capital issued Information Technology (3297.8 Million)
- Communication (3064.1 Million)
- Capital Investment Information Technology (4495.2 Million)
- Communication (5925.6 Million)
- Labor Force (IT Companies) Information Technology (33669 Job opportunities)
- Communication (5056 Job opportunities)
- Telecom Egypt Investment (3001 Million)

Human Resources

Skills Development

Number of trainees (113732 Thousands)

Professional Training

Number of trainees (22588)

IT Clubs

Number of IT clubs (899)

Table 16.1: Telecommunications Infrastructure Indicators

Detail	*Indicator*	*October 1999*	*March 2004*	*March 2005*
Fixed Services	Exchange capacity	6.4 million	11.6 million	12.1 million
	No. of fixed telephone line subscribers	4.9 million	9.1 million	9.6 million
	No. of exchanges in rural areas	775	1082	1112
	No. of public phone cabins	13,300	48,983	54,346
Mobile Lines	No. of mobile phone subscribers	654,000	6.1 million	8.6 million
	No. of mobile phones per 100 people	0.98	8.9	12.3
	Mobile service companies	2	2	2
Internet Penetration	Internet capacity	20 Mbps	1.1 Gbps	2.0 Gbps
	No. of internet users	300,000	3.15 million	4 million
	Monthly internet subscription	100 LE (29 US$)	Cost of local call	Cost of local call
PC Penetration	No. of PCs	850,000	1.7 million	2.0 million
	No. of PCs per 100 people	1.25	2.5	2.85

Table 16.2: Private Sector Participation in the Communication Sector

Service Type	*October 1999*	*June 2005*
Internet service providers	40	214 - 303 number
Mobile	2	2
Public data network	1	8
Internet backbone providers	1	4
Value added voice services	—	2
Public phone services	3	3
Equipment manufacturing	2	3
Copper wire manufacturing	3	4
Fiber optic cables	2	2
VSAT	—	2
GMPCS	—	2

Egypt's telecommunications infrastructure has increased dramatically during the last few years. Telecom Egypt has revolutionized its service portfolio, and the developments in delivering access to the Internet have been colossal, and mostly implemented through public private partnership schemes. Egypt's mobile phone operators complement a rapidly advancing telecommunication industry.

The use of wireless technology and its developments has been seen to be a decisive field in an effort to provide universal access, particularly to outlying rural areas.

E-Access

A basic requirement for the development of a true information society is that its citizen have access to information and communications technology. The e-access schemes set out to guarantee universal, easy, affordable and rapid access for all its citizens to ICT, and stimulating awareness of the potential uses and benefits of ICT.

The invitation is being carried out through public-private partnerships and aims to create an investment climate that encourages private sector investment in the development of infrastructure and provision of services.

Schemes detailed are the PC for Every Home, Free Internet, IT clubs and the Broadband Initiative.

E-Government

The main drive of e-government is to modernise the citizen's experience of public services and to improve the functionality of the Egyptian government by increasing the ease with which citizens can access services and bring greater efficiency to government operations.

E-Business

The e-business initiative is to build and expand a strong information and technological infrastructure that contributes to supporting the business community and encouraging direct foreign investment.

E-Health

The initiative intends to synchronise efforts between health authorities and ICT facilities and to increase use and availability of patients' medical history for medical practitioners and hospitals by different media.

E-Content

The scheme will promote the development of materials in a wide range of categories including books, media, arts, education, sports and entertainment. The initiative is designed to enrich Arabic content in digital format and to encourage wide ranging demand for such material and promote products and services offered.

E-Culture

A national plan is being designed and implemented in managing and protecting cultural heritage, promoting cultural contents, supporting artistic creation and new ways of expression, and reinforcing and cultivating the Arabic language. The plan includes measures to improve the skills and efficiency of units working in the field of cultural and natural heritage by introducing automation concepts and IT in documentation and publishing.

PC for Every Home

The goal of making the Internet affordable can be achieved only through the introduction of affordable access devices. To increase PC usage and to attain the level of penetration to reach one PC per every three families, the government is working in partnership with the private sector to develop and initiative to provide the hardware needed to get more people online. The aim of the PC for Every Home initiative is to provide end- users with computers at prices affordable for the average user. By the end of 2007, fourteen million users are expected to benefit from four million PCs.

All PCs are being sold with a pre-installed Microsoft Windows-XP Operating system. Locally developed value added software is being made available with the PCs. The credit history of the fixed phone subscriber is checked and the PC is guaranteed by the fixed phone line. By March 2005 the number of PCs sold in this manner was 66,000. The target is to reach 100,000 PCs sold annually to lower market segments and to open the market to Egyptian and foreign investors.

Free Internet

In January 2002, the government's first major success in its effort to make technology more affordable came with the launching of the Free Internet Initiative in Cairo. The free Internet model is a private partnership, a joint effort between MCIT and Telecom Egypt in co-operation with the majority of Egypt's private Internet Service Providers (ISPs). The initiative offers subscription free Internet services to Internet users via dial up to special –prefix numbers starting with '0777' or '0707'. In September 2002, free Internet Service was available nation wide. Today, Internet users across Egypt are only charged for the price of local phone calls associated with connecting to the network. The local phone call charges are approximately 0.15 US$ for an hour of access. The revenues from the free Internet calls are shared between Telecom Egypt and the service providers. The free Internet model was a major step in increasing the number of Internet numbers.

The program effectively abolishes the major cost barrier of Internet access for users and has encouraged many who had previously not made use of web to log on. After introducing the free Internet model, the number of Internet users more than doubled, increasing from one million in January 2002 to 4 million in March 2005 through more than one million connected households.

Broadband

Broadband has been growing worldwide at a very high rate compared to other telecommunication services. Governments all across the world are more and more emphasizing its importance for economic, social and cultural development. Egypt remains firmly committed to its goal of bridging the Digital Divide. The government makes stern efforts to ensure that Egypt derives the maximum benefit from information and communication technology. This commitment to fostering a transition to an Information Society is made not as an end in itself but rather as a means to the more fundamental goals of the Egyptian government. These include improved public services, stronger economy, increased productivity and greater opportunities for all.

In line with its stated vision, Egypt is leading an ambitious broadband initiative, with the objective of increasing broadband penetration. The initiative aims at providing all its citizens with easy and affordable access to the opportunities offered by new technologies. Started in May 2004, the Broadband initiative is promising affordable high-speed connection to the Internet. Collaborative efforts from the government and the private sector are playing an important role in achieving these objectives and spreading broadband services in Egypt.

Increasing ADSL penetration comes first on the broadband initiative agenda with an initial target of 50,000 subscribers during the first year. To meet this target, the initiative is focusing during its initial phase on three main numerical items:

- Decreasing monthly charges for 256 kpbs ADSL services by 50% to reach 150 LE:
- Reducing ADSL installation procedure to range from a few days to a week.
- Increasing ADSL public awareness through vigorous marketing campaigns.

The scheme targets primarily residential users as well as small enterprises. Statistics to date indicate an increase in the number of ADSL subscribers from eight thousand back in May 2004 to exceed thirty three thousand subscribers today. The end user is promised an improved means of Internet connectivity, by offering high-speed connection, a constant availability of the phone line to receive voice calls irrespectively, and the possibility for more than one PC capital to share the same Internet connection. The Broadband initiative is expected to drive the development of local electronic content and online applications in the various sectors of the economy.

Innovation

More emphasis is being put on innovation the associated building of Intellectual Property and the consequent emerging number of Start Ups. As a public private partnership the ICT technology development fund was established to invest venture capital in Egyptian Start Ups. With a combination of an enabling economic environment, a sustained effort towards main streaming information technologies, and a dynamic education system encouraging innovation, the government is confident real steps forward can be mad in the near the future to an all inclusive information society.

Innovation realize on an extensive program of research and development, revolving around areas of Data Mining and Computer Modelling, Wireless Communication, Mobile and E-Services, and Electronic Design.

Particular attention now is being paid to the formation of the incubators, intending to provide support and a strengthening to existing companies, as well as attracting multinational companies to invest in developmental activities in Egypt.

Establishment of the CIT Research and Development Centres of Excellence is now a crucial step in building an environment where ICT innovation and excellence can thrive. It is hoped they will provide leadership and guidance for transforming the creation and commercialisation of ICT to a new level as well as facilitate the introduction of world class ICT research that will raise the standard of Research and Development in Egypt. The process is hoped to offer a magnet to attract world-class researchers to Egypt including many of Egypt's best and brightest researchers who are currently working overseas.

The purpose of these centres is to provide a platform to performing the research and development (R&D) conducted by Egyptian experts in their respective fields of specialization. Such R&D can be performed in favour of local as well as multinational institutions. This activity will also help create Start Up companies, specialized in domains of these centres. These centres of excellence shall also play the role of ' pre-incubator', as explained in the next section of the report.

The centres will be created in the form of consortiums consisting of leading local and multinational industrial organizations operating in Egypt. In addition to this, both private and public collaborate with Egyptian universities and research institutions. One of such universities or research institutes will act as the lead on a given centre and focal point for co-ordination with MCIT.

The later will act as a catalyst by helping bring the various participants to the consortium, by giving its input about local as well as world market conditions in the domain of CIT, and by supporting the establishment of the centres by cash or 'in kind' contributions.

In the initial phase, four centres will be established in the following focused areas.

- Data Mining and Computer Modelling
- Wireless Communication
- Mobile and E- services
- Electronic Design

Source: Egypt's information society, a publication of the ministry of communication and information technology.

Electronics Research Institute and it's Role for Supporting the Industry

ERI started its activities in 1963 as the electrical and electronics engineering laboratory at the National Research Centre, NRC. In 1989 a decree was issued to establish the Electronics Research Institute affiliated to the Ministry of Scientific Research.

ERI was established with the vision to be a centre of excellence in the field of electronics and informatics by national, regional and international standards.

The mission of ERI is to undertake basic and applied research in advanced technologies to enhance economic, industrial and strategic competitiveness through increased value added quality and productivity.

ERI is composed of the following departments:

1. Power electronics and energy conversion
2. Computers and systems
3. Informatics
4. Microstrip circuits
5. Microwave engineering
6. Photovoltaic Cells
7. Microelectronics

Table 16.3 shows some projects and activities where ERI is involved.

On going Research Projects at ERI, Microelectronics Dept.

RF Circuits

- Dual-loop PLL based Frequency Synthesizer for multistandard Transceivers (Ph. D project)
- Distortion analysis of RF circuits (M. SC project)
- Design and Implementation of RF MEMS Components for Wireless Applications (NSF project with USA)
- Design and Fabrication of RF-SAW Sensor Using MEMS Technology for wireless Applications (NSF project with USA)

Digital Circuits

- A new adaptive approach for signal compression using Wavelets (Ph.D project)
- VLSI implementations of fast division algorithms (M. SC project)
- Automatic modulation recognition of communication circuits (M. SC project)

MEMS

- Piezoelectric cantilever microphone for hearing aid devices (M. SC Project)

Table 16.3

Name of Research/Technology Development Body	Type		Location		Nature of Cooperation/Association/Agreement
	Local	Intl.	Country	Town	
Academy of Scientific Research & Technology	X		Egypt	Cairo	Research Project: Research In the Satellite Receiving Equipment including Nile SAT
Academy of Scientific Research & Technology	X		Egypt	Cairo	Research Project: Development fo Human Resources in the Field of Advance and New Technology (Electronics, Computers and Informatics
Academy of Scientific Research & Technology	X		Egypt	Cairo	Research Projec: Studies for Design and Production Automation Especially in the Field of Robots
	X		Italy	Palermo	Research Project: Identification and Control of Electric Drives and Systems
University of Salford	X		UK	Salford	UK Link Project : Advanced Teaching and Training in the Use of Geographical Information Systems (GIS) for Environment Planning and Healthcare Support
University of Hawaii	X		USA	Hawaii	Workshop on Recent Advances in Wireless Telecommunications Technologies
University of Louisiana	X		USA	Lafayette, LA	US-Egypt Joint Fund Project: A Low Cost Video and Image Compression Framework
Ohio State University	X		USA	Colombus, OH	Research Project : Design and Implementation of cts for the Third Generation Mobiles
Ohio State University	X		USA	Colombus, OH	Research Project : Design of RF MEMS Components for Wireless Applications
Central Electronics Engineering Research Institute (CEERI)	X		India	Pilani	Collaboration in Establishing Microfabrication Facility
Tecnopolis CSATA Novous Ortus	X		Italy	Bari	Cooperation to build a VLSI Design Support Center at ERI
Academy of Scientific Research & Technology (STC Program)	X		Egypt	Cairo	Research Project: Computer Control in Textile Industry (Implemented in STIA)

Contd...

Table 16.3–Contd...

Name of Research/Technology Development Body	Type		Location		Nature of Cooperation/Association/Agreement
	Local	Intl.	Country	Town	
Academy of Scientific Research & Technology (STC Program)	X		Egypt	Cairo	Research Project: Improving the Jacquard Looms in the Oriental Weavers Co.
Academy of Scientific Research & Technology (STC Program)	X		Egypt	Cairo	Research Project: Computer Control in EDFINA Co.
Academy of Scientific Research & Technology (STC Program)	X		Egypt	Cairo	Research Project: Computer Based Quality Improvement for the Products of the International Electronics Co.
Academy of Scientific Research & Technology (STC Program)	X		Egypt	Cairo	Research Project: Use of CAD/CAM in Textile Industry (Implemented in Kafr El-Dawar Co.)
TELIMISR Co.	X		Egypt	Cairo	Design and Fabrication of Front End Circuit for TV Receiving From Satellite
BALAEEM Petroleum Co.	X		Egypt	Eastern Desert	Adjusting the Communication System of the Company between its Headquarters and the Sector in the Eastern Desert
Academy of Scientific Research & Technology (STC Program)	X		Egypt	Cairo	Research Project: Study for the Development of the Electronics Industry in Egypt and Suggesting Projects for its Development
Electronics Facotry, Arab Organization for Industrialization	X		Egypt	Cairo	Agreement for Cooperation in Design and Fabrication for RF cts
Social Fund for Development (Consultation Unit)	X		Egypt	Cairo	Agreement for HRD and providing technical Assistance in proposing projects for SME's in the fields of Electronics, IT, and Communications
Technical University of Athens (within the framework of the EU)	X		Greece	Athens	Research Project: Using HPC for Quality Control in Textile Industry
Technical University in the Netherlands (within the framework of the EU)	X		Netherlands		Research Project: DINAR an Automatic Translation Tool
IDEAL Company	X		Egypt	Cairo	Design and testing Lab. For 1-phase Induction Motors
Ministry of Education	X		Egypt	Cairo	Research Project: Development of Educational Programs for Handicapped Children

17

Leveraging Bioinformatics for Electronic Sensor Based Biological Micro-array Chip Design

Arif Anwar
Vice President,
Synamatix Sdn. Bhd., B.Sc. (University of London), Ph.D. (University of Oxford, London)
E-mail: arifanwar@symantix.com

Agenda

- Why do we need microarrays?
- What is a microarray?
- Where is the technology today?
- What are the problems?
- Why do we need Bio-MEM?
- What part does (Synamatix) bio-informatics play?

Gene Expression

- All of the cells in our body have the same DNA
- Why are they different from each other?
- Each type of cell expresses a different set of genes
- Two cells that express the same gene may express them at different levels.

Goals of Gene Expression Studies

- Make a guess at gene function – "guilt by association"
- Genes with similar patterns of expression may be in similar pathways
- Distinguish cancer cells from normal cells on the molecular level (to identify diagnostic or prognostic markers, or to identify drug targets)
- Characterize complex diseases at the molecular level

Measuring Enable 100000 + Northern Blots per Experiment

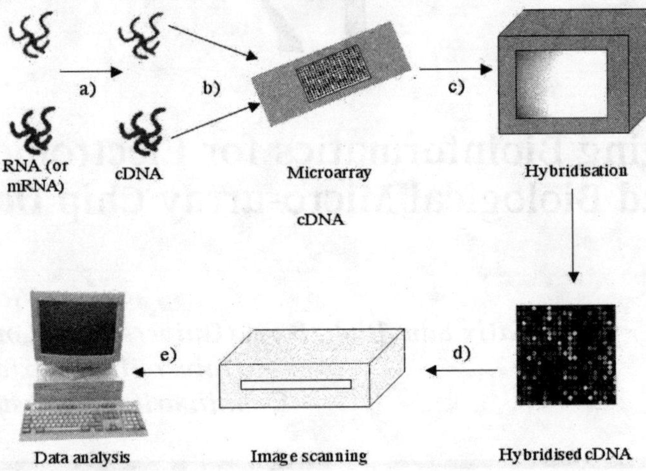

Problems

- Imprecise science
- Wet
- Bioinformatics often wrong
- Difficult to analyse results
- Need to generate cross correlations
- Johns Hopkins study

Comparison of Different MEM DNA Detection Systems

Type of sensor	Advantages	Disadvantages
Direct DNA electrochemistry	Highly sensitive (femtomoles of target); requires no labeling step; amenable to a range of electrodes	High background signals; cannot be multiplexed; destroys the sample
Indirect DNA electrochemistry	Highly sensitive (attomoles of target); usually requires no labeling step; multiple-target detection at same electrode	Probe substrate can be difficult to prepare; destroys the sample
DNA-specific redox indicator detection	Moderate to high sensitivity (femtomoles of target); well suited to multiple-target detection; samples remain unaltered	Chemical labeling step required unless 'sandwich' method used; sequence variations can be problematic
Nanoparticle-based electrochemistry amplification	Extremely sensitive (femtomole to zeptomole range, 10^{-15} to 10^{-21} moles); well suited to multiple-target detection with different nanoparticles	Many development steps in assay; reliability and robustness of surface structures problematic; sample usually destroyed
DNA-mediated charge transport	Highly sensitive (femtomole range) and simple assay; requires no labeling; uniquely well suited for mismatch detection; sequence independent; amenable to multiplexing; applicable to DNA-protein sensing step	Biochemical preparation of target sample required

Electronic Detection of DNA by its Intrinsic Molecular Charge

- Selective and real-time detection of label-free DNA using an electronic readout.
- Microfabricated silicon field-effect sensors used to monitor increase in surface charge when DNA hybridizes on sensor surface.
- Electrostatic immobilization of probe DNA on a positively charged poly-l-lysine layer allows hybridization.
- Nanomolar DNA concentrations can be detected within minutes, and a single base mismatch within 12-mer oligonucleotides can be distinguished.
- The sensors were fabricated by standard silicon microtechnology.

Commercial–Field Programmable/Electronic Microarrays

- Nanogen has produced a silicon chip embedded with 100 "programmable" probe pads
  - ❖ *80mm platinum pads (each spaced about 200 um apart)*
  - ❖ *Each pad can have apply a voltage (-1.3 to 2.0 V)*
- Since DNA carries a negative charge, applying a positive charge on a pad "corrals" DNA onto that spot
  - ❖ *This is used to build custom arrays by washing the chip in a single stranded DNA solution, biasing the desired spot on the chip, and then chemically fixing the DNA to that spot*
- The electric charge is also useful during the hybridization reaction
  - ❖ *Pooling the DNA onto the charged pads increases the reaction by a factor of 1000*
  - ❖ *Reversing the charge "shakes loose" imperfectly matched DNA leading to more accurate results*

Bio MEM

- Electronic microarrays from Nanogen contain 100 test sites laid out in a geometric grid.
- Each test site on the Nano Chip array can be controlled electronically from the Nano Chip Workstation's on-board computer.
- A permeation layer coated on the chip acts as a protective interface between the electrically active surface and the biological test environment.

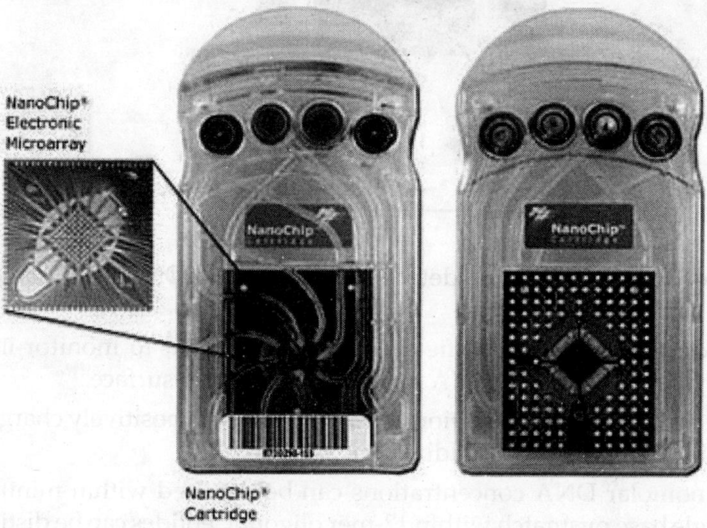

A Closer Look

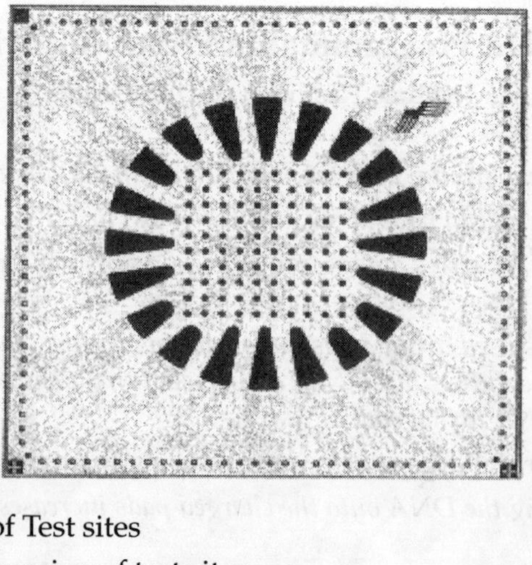

- 10 X 10 array of Test sites
- Electronic addressing of test sites
- Molecule migration based on field charges

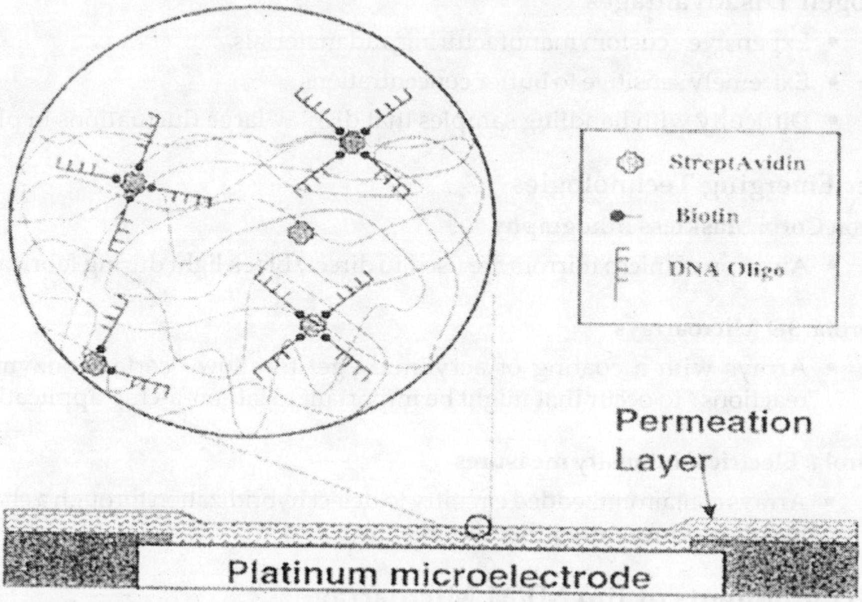

- Controlled position of capture probes / query samples
- Charge migration of samples in test site
- Affinity Binding
- *Avidin - matrix ; Biotin - sample*

Applications

- Determination of new genomic sequences.
- Scanning of genes for polymorphisms that might have an impact on phenotype
- Wholesale survey of the pattern of gene(s) expression in organisms when exposed to particular nutrient or chemical (or physical) insult.
- Multiple signal processing (multiplex) capabilities on a signal chip
- Multiple experimental investigations on 1 chip platform
- Apoliprotein (ApoE) and ApoE allele specific analysis, snp discrimination and isoform detection

Advantages

Normal Microarrays	*Field-Programmable Microarrays*
Limited to 1 experiment per chip	Perform multiple experiments on a single chip
Passive hybridisation	Electronically programmed hybridisation
High nonspecific binding	Charged environment allows concentration of samples up to 10 fold

Nanogen Disadvantages

- Expensive - custom manufacturing and materials
- Extremely sensitive to buffer concentrations
- Difficulty with handling samples that display large fluctuations in pH

Other Emerging Technologies

Xeotron Corp: Maskless lithography

- An array of micro mirrors are used to direct/block light during fabrication

Motorola: 3D Microarrays

- Arrays with a coating of acrylimide gel to allow "certain enzymatic reactions" to occur that might be important to lab-on-a-chip applications

Motorola: Electrical intensity measures

- Arrays contain embedded circuitry to detect hybridization through a change in conductance rather than fluorescence

Key components of Bio MEM Microarray

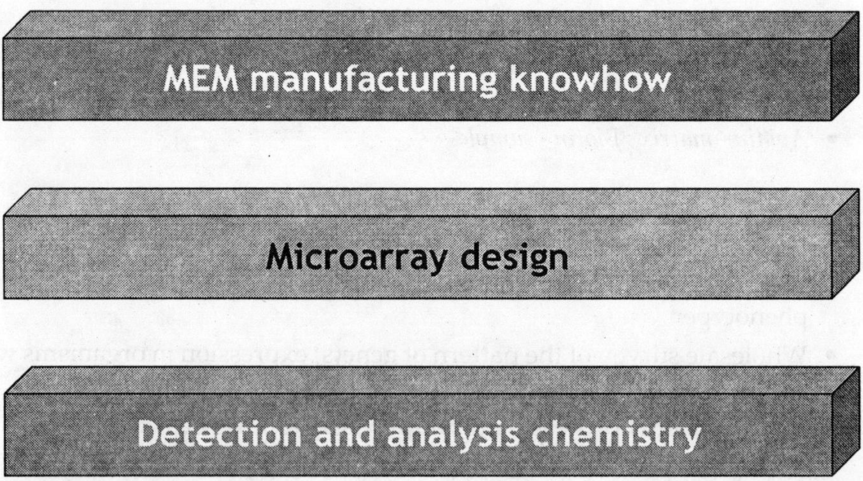

Synamatix Background

- Current database platforms will not be able to scale to manage ever increasing data volume and complexity
- Novel database platform to meet needs, not a:
  - ❖ Suffix tree
  - ❖ Relational database
  - ❖ Suffix array

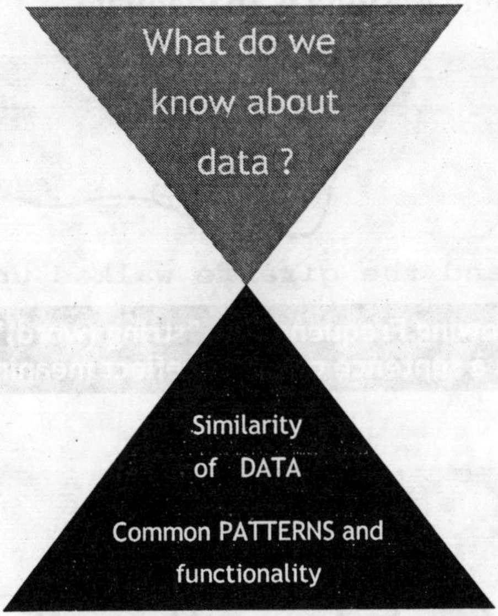

- There is no evolutionary need to preserve non-functional sequence patterns
- Evolution requires the conservation of patterns which are at least functionally equivalent, or functionally better
- Significant patterns and their relationships are extended and maintained by SynaBASE

3 Ways to Access to SynaBASE and Applications

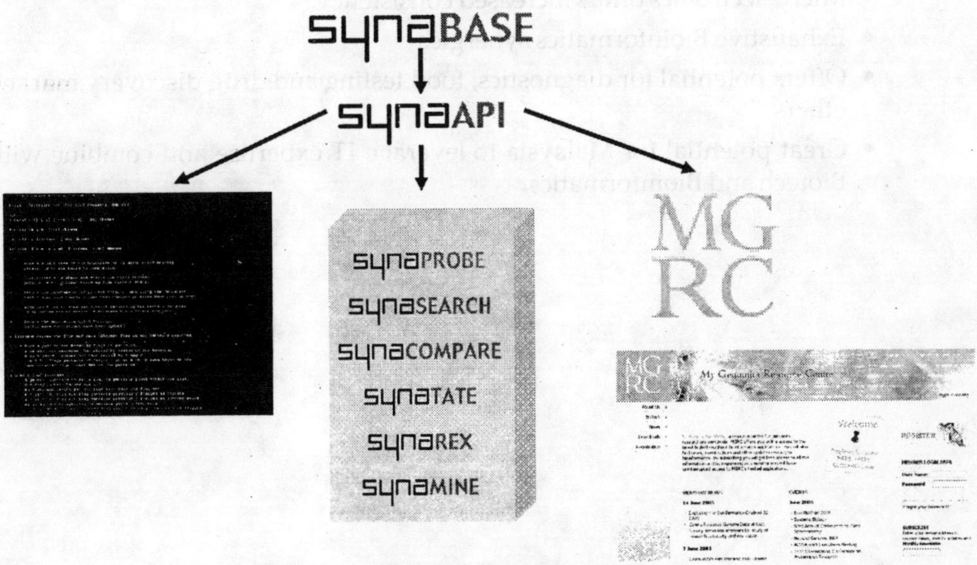

Relying on Frequency Alone is Inadequate...

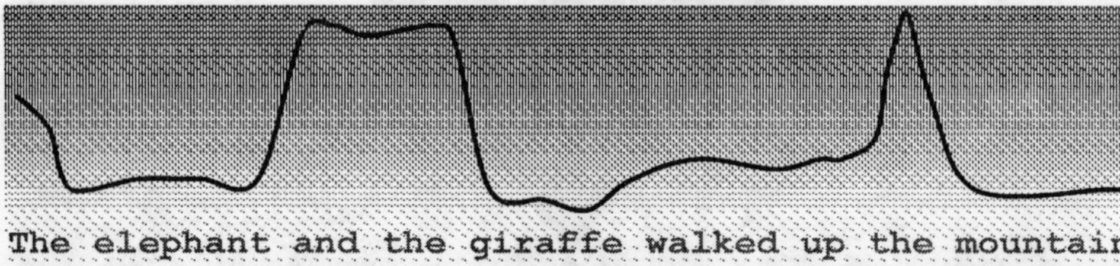

A graph showing Frequency of "string (word)" patterns in a sentence does not reflect meaning

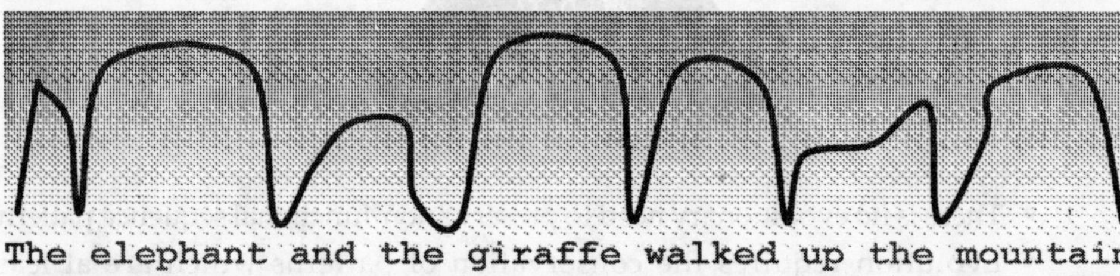

A graph showing Probabilities of predicting Precessor and Successor Characters/events (string Significance) reflecting meaning

Summary

- Microelectronics offers increased consistency
- Exhaustive Bioinformatics synergies
- Offers potential for diagnostics, food testing and drug discovery market, others
- Great potential for Malaysia to leverage IT expertise and combine with Biotech and Bioinformatics.